Lecture Notes in Mathematics

Edited by A. Dold and B. Eckmann

326

Alain Robert

Elliptic Curves

Notes from Postgraduate Lectures Given
in Lausanne 1971/72

Springer-Verlag
Berlin Heidelberg New York Tokyo

Author

Alain Robert
Université de Neuchâtel, Institut de Mathématiques
Chantemerle 20, 2000 Neuchâtel, Switzerland

1st Edition 1973
2nd Corrected Printing 1986

Mathematics Subject Classification (1980): 12B35, 12B37, 14G10, 14H15, 14H45, 32G15

ISBN 3-540-06309-9 Springer-Verlag Berlin Heidelberg New York Tokyo
ISBN 0-387-06309-9 Springer-Verlag New York Heidelberg Berlin Tokyo

2146/3140-543210

Notations and Conventions

We have used the usual letters for the basic sets of numbers : $\mathbb{N}$ (natural integers 0,1,2,...), $\mathbb{Z}$ (ring of rational integers), $\mathbb{Q}$ (field of rational numbers), $\mathbb{R}$ (field of real numbers), $\mathbb{C}$ (field of complex numbers), $\mathbb{F}_q$ (finite field with q elements). As a rule, we denote by $A^{\times}$ the multiplicative group of units (invertible elements) in a ring A. In formulas, the cypher 1 always represents the number one (except in log x ... so that in one occurence I have used log -1 to avoid ambiguities). Also $\underline{e}(x) = e^{2\pi i x}$ (normalized exponential).

In a theorem, I list properties under Latin letters a), b), ... keeping i), ii), ... for equivalent properties, but the meaning is always clear by the context.

The following system has been adopted for cross-references. All theorems, propositions, corollaries, lemmas, remarks, definitions, formulas, errata,... are numbered in one sequence. Such a cypher as (III.3.4) refers to the item (3.4) of chapter III, i.e. the fourth numbered in section 3. (This happens to be a lemma 3.) From inside chapter III we would refer to (3.4) (in section 3, sometimes simply to lemma 3 : this last system of numeration has not been used systematically, but only when it can be more suggestive locally).

ERRATA

Pages numbers refer to bottom pages numbers.

p.73 line 2 from bottom. It is not true that $\tau(n)$ is divisible by 691 for nearly all integers. For example, when p is prime, $\tau(p) \equiv 1 + p^{11}$ mod 691 and the set of primes p such that $\tau(p)$ is divisible by 691 has density 1/690 according to Dirichlet's theorem. But $\tau(n) = \sigma_{11}(n)$ mod 691 and $\sigma_{11}(n)$ is divisible by 691 for "nearly all n" in the sense of the naïve density on the set of integers... cf. J.-P. Serre, C.R.Acad.Sc.Paris 279 (1974), pp.679-682.

p.74 Since Weil's conjectures have been proved by P. Deligne, Ramanujan's conjecture is true.

p.102 In the case of a double point, the isomorphism of the smooth part of the singular cubic with the multiplicative group is defined over an extension of degree $\leqslant 2$ of the base field : it is enough to take the extension generated by the two slopes at the singular point (these are quadratic numbers).

p.183 Demjanenko's claim of proof has however remained unverified so that the strong conjecture is still considered as open.

p.231 line 4. The proof of part b is incorrect. A correct argument is to be found in Séminaire Chevalley 1958/59: Variétés de Picard p.6.06-6.08 .

p.234 line 6. the numbers $\binom{-\frac{1}{2}}{k}$ are not in $p\mathbb{Z}_p$ when $k > \frac{1}{2}(p-1)$ and the hypergeometric series $F(\frac{1}{2},\frac{1}{2}:1:\lambda)$ has to be truncated. The reduction mod p of the series of $F(\frac{1}{2},\frac{1}{2}:1:\lambda)/F(\frac{1}{2},\frac{1}{2}:1:\lambda^p)$ is a finite series congruent to Hasse's polynomial (up to sign). This still makes it clear that the Hasse polynomial is a solution of the hypergeometric differential equation. cf. B. Dwork, Publ.Math. I.H.E.S 37 (1969) pp.47-49.

A.R. Feb.'85

P.S. Correct spelling: Bernoulli (pp.66-67)
Hurwitz (p.256)

Table of Contents

INTRODUCTION

Elliptic curves are special cases of two theories, namely the theory of Riemann surfaces (or algebraic curves) and the theory of abelian varieties, so that any book concerned with these more general topics will cover elliptic curves as example. However, in a series of lectures, it seemed preferable to me to have a more limited scope and introduce students to both theories by giving them relevant theorems in their simplest case. I think that the recent recrudescence of popularity of elliptic curves amply justifies this point of view. I have not chosen the most concise style possible and sometimes have committed the "crime of lèse-Bourbaki" by giving several proofs of one theorem, illustrating different methods or point of views.

I shall not give here any idea of the topics covered by these notes, because each chapter has its own introduction for that purpose (prerequisites are also listed there). Let me just mention that I have omitted complex multiplication theory for lack of time (only integrality of singular invariants is proved in chapter III). In the short commented bibliography (given for each chapter at the end of the notes), I quote most of my sources and indicate some books and articles which should provide ample material for anyone looking for further reading.

The origin of my interest in elliptic curves has to be traced to a series of lectures given by M. Demazure (Paris Orsay, oct.-dec. 67) on elliptic curves over $\mathbb{C}$. Although the presentation I have adopted differs somewhat from his, I have been much influenced by the notes from these lectures (especially in the section on theta functions). I would also like to seize the opportunity of thanking here Y. Ihara,

J.-P. Serre, G. Shimura for very helpful discussions, correspondence... Only at the end of the lectures did I learn through S. Lang that he had also written a book on elliptic curves. It seemed however that the material covered was sufficiently different to allow the publication of my notes, and I hope that they will still have some use.

Finally it is a pleasure to thank the audience of the lectures whose interest stimulated me, my wife who gave me some hints on language and L.-O. Pochon who proof-read most of the notes, pointed out some mistakes and established an index. However, I take full responsability for remaining mistakes and would be grateful to anyone bothering to let me know about them !

September 1972

A. Robert
Institut de Mathématiques
Université de Neuchâtel
CH-2000 NEUCHATEL
(Switzerland)

CHAPTER ONE

Complex Elliptic Curves

This chapter has as first aim the presentation of the classical theory of elliptic functions and curves, as first studied in the nineteenth century by

Abel, Jacobi, Legendre, Weierstrass .

In particular, the following mathematical objects will be shown to be equivalent :

i) Compact complex Lie group of dimension one.

ii) Complex torus $\mathbb{C}/L$, with a lattice $L \subset \mathbb{C}$.

iii) Riemann surface of genus one (with chosen base point).

iv) Non-singular plane cubic of equation

$$y^2 = x^3 + Ax + B \quad .$$

Here, we shall basically only assume that the reader is familiar with the theory of analytic functions of one complex variable. However, we shall mention briefly some "superior" interpretations which can be fully understood only with some knowledge of the basic definitions and properties of vector bundles and sheaves on Riemann surfaces.

1. Weierstrass' theory

A lattice in a (finite dimensional) real vector space V is by definition a subgroup L generated by a basis of V. Thus a lattice is isomorphic to a group $\mathbb{Z}^d$ where d = dim(V), and V/L is a compact group. When we speak of lattice in a complex vector space, we always mean lattice for the real vector space of double dimension obtained by restricting the scalars from $\mathbb{C}$ to $\mathbb{R}$. Thus a lattice L in $\mathbb{C}$ is a subgroup of the form $L = \mathbb{Z}\omega_1 + \mathbb{Z}\omega_2$ with two complex numbers ω_i such that ω_1/ω_2 is not real.

Let L be a lattice in $\mathbb{C}$ which we shall keep fixed. We say that a meromorphic function f on $\mathbb{C}$ is an elliptic function (with respect to L, or L-elliptic) when it satisfies

$$f(z + \omega) = f(z) \quad \text{for all } \omega \in L$$

(whenever one member is defined!). An elliptic function can be considered as a meromorphic function on the analytic space $\mathbb{C}/L$. It is obvious that the set of elliptic functions (with respect to L) is a field with respect to pointwise addition and multiplication. It is nonetheless obvious that if f is elliptic, so is its derivative f' .

Let $f \neq 0$ be a non identically zero elliptic function. For any point $a \in \mathbb{C}$, we can use the Laurent expansion of f at a to define the rational integer $\mathrm{ord}_a(f)$ (smallest index with non-zero Laurent coefficient). This number is positive if f is regular at a, and strictly negative if f has a pole at a. By periodicity of f, we get $\mathrm{ord}_{a+\omega}(f) = \mathrm{ord}_a(f)$ for any $\omega \in L$, hence we shall often consider a as an element of $\mathbb{C}/L$ to speak of $\mathrm{ord}_a(f)$. The formal sum of elements of $\mathbb{C}/L$

$$\mathrm{div}(f) = \sum \mathrm{ord}_a(f)\cdot(a) \ ,$$

has only finitely many non-zero coefficients (because $\mathbb{C}/L$ is compact and the zeros and poles of f are isolated), hence can be considered as element of $\mathrm{Div}(\mathbb{C}/L) = \mathbb{Z}[\mathbb{C}/L]$ (we consider this set as additive group, without the convolution product). The first algebraic properties of elliptic functions are given by the following

(1.1) Theorem 1. Let f be a non-constant L-elliptic function, and

$$\mathrm{div}(f) = \sum n_i(a_i) .$$

Then we have:

a) f is not holomorphic.

b) $\sum \mathrm{Res}_{a_i}(f) = 0$.

c) $\sum n_i = 0$.

d) $\sum n_i a_i = 0$ in $\mathbb{C}/L$.

Proof: If an elliptic function f is holomorphic, it is bounded in a period parallelogram (this is a compact set), hence bounded in the whole complex plane by periodicity. It must then be constant by Liouville's theorem, hence a). We choose now a basis ω_1 , ω_2 of L and $a \in \mathbb{C}$ such that f has no zero or pole on the sides of the period parallelogram P_a of vertices a, $a + \omega_1$, $a + \omega_2$, $a + \omega_1 + \omega_2$. The integration of $f(z)dz$ on the (anticlockwise) oriented boundary dP_a of P_a gives $2\pi i \sum \mathrm{Res}_{a_i}(f) = 0$ because the opposite sides cancel their contributions by periodicity of f. This proves b). Then c) follows from b) applied to the elliptic function f'/f . For d), we integrate the function zf'/f over the same boundary dP_a . The contributions from parallel sides do not cancel here, but e.g.

$$\left(\int_a^{a+\omega_1} - \int_{a+\omega_2}^{a+\omega_1+\omega_2}\right) zf'(z)/f(z)dz = -\omega_2 \int_a^{a+\omega_1} f'/f\, dz .$$

Since f admits the period ω_1 , $f([a,a+\omega_1])$ is a closed curve C and

$$\int_a^{a+\omega_1} f'/f\, dz = \oint_C \frac{d\zeta}{\zeta} = 2\pi i \cdot \mathrm{Index}(0;C) = 2\pi i m_1$$

Adding these two similar terms gives $\sum n_i a_i = m_2\omega_1 - m_1\omega_2 \in L$ q.e.d.

We make a few comments on the meaning of this theorem. First, a) asserts that a non-constant elliptic function f has at least one pole. Moreover, if f and g are two elliptic functions with same divisor, the quotient f/g has no pole, hence is a constant c : $f = c \cdot g$. This shows that the divisor of an elliptic function determines the function up to a multiplicative constant. Then b) shows that an elliptic function f cannot have only one pole, this pole being simple. An elliptic function can have only one double pole with zero residue, or two simple poles with opposite residues. These are the simplest possibilities. Then c) brings out the fact that an elliptic function has as many zeros as poles. Thus if f is any non constant elliptic function and c <u>any</u> complex value, f and f - c having the same poles must have the same number of zeros. This explains that f takes all complex values the same number of times. This constant number of times f (non-constant) assumes every value may be called the <u>valence</u> (or the <u>order</u>) of f. For any divisor $\sum n_i(a_i) \in \mathrm{Div}(E)$, where we put $E = \mathbb{C}/L$ to simplify the notations, we define its <u>degree</u> as being the sum $\sum n_i$. This gives a group homomorphism

$$\deg : \mathrm{Div}(E) \longrightarrow \mathbb{Z} .$$

With this definition, c) can be restated as $\deg(\mathrm{div}(f)) = 0$ for all elliptic functions $f \neq 0$. Finally d) gives a further necessary condition on the divisor of an elliptic function (Abel's condition). We shall prove that the conditions c) and d) on a divisor of E are also sufficient to assure the existence of an elliptic function having precisely that divisor.

We have to prove the existence of non-constant L-elliptic functions, and we shall follow Weierstrass' idea of constructing an elliptic function with only a double pole (on E), with zero residue. The construction is of transcendental nature.

For that purpose, we try an expression of the form $\sum(z - \omega)^{-2}$, provided it converges. But for z fixed, and $|\omega| \to \infty$, $(z - \omega)^{-2} \sim \omega^{-2}$.

(1.2) Lemma. *Let* V *be a (finite dimensional) real Euclidian vector space, and* L *a lattice in* V. *Then* $\sum_{0 \neq \omega \in L} \|\omega\|^{-s}$ *converges absolutely for* $\mathrm{Re}(s) > \dim(V)$.

Proof. We put d = dim(V) and we choose a basis ω_1 , ... , ω_d of L. We observe first that it is sufficient to prove the convergence of the sum extended over the $\omega = \sum n_i \omega_i$ with positive $n_i \geqslant 0$, not all zero. We group together those for which $\sum n_i = n > 0$ is a given strictly positive integer, and choose two constants $A > 0$, $B > 0$ such that

$$A/n \leqslant \frac{1}{\|\omega\|} \leqslant B/n \quad .$$

The number of these ω is asymptotically equal to the area of the simplex in V with vertices $n\omega_i$, hence is of the order of Cn^{d-1} $(C > 0)$. (Actually the exact number can be computed, it is the binomial coefficient $\binom{n+d-2}{d-1}$.) We find thus a contribution to the sum of the order of n^{d-1}/n^s. The absolute convergence of this series occurs exactly when $\mathrm{Re}(s) - d + 1 > 1$, hence for $\mathrm{Re}(s) > d$ as asserted.

Now, the series of general term $(z - \omega)^{-2}$ does not converge, but if we only subtract the asymptotic behavior ω^{-2}, we get terms $(z - \omega)^{-2} - \omega^{-2}$ which must be of the order $c(z)\omega^{-3}$ (the incredulous reader can check that by computing the difference of these fractions!), hence will give a convergent series. We define the Weierstrass' function

$$\wp(z) = \wp(z:L) = z^{-2} + \sum{}'\{(z - \omega)^{-2} - \omega^{-2}\} \quad , \tag{1.3}$$

the sum being extended over $\omega \in L' = L - \{0\}$, which we indicate by a prime in the summation (as a rule, we shall omit the prime for sums extended over all L, and keep it for sums over L'). This series is now absolutely convergent (for all $z \notin L$), and it is uniformly (and normally) convergent when z stays in a bounded part of $\mathbb{C} - L$.

It is not absolutely obvious that $\wp$ admits L as lattice of periods, but it is an even function, and we proceed as follows. The derivative of $\wp$ can be computed by termwise differentiation :

(1.4) $$\wp'(z) = \wp'(z:L) = -2\sum (z - \omega)^{-3}$$

(no convergence factors are needed here by the lemma, and they have disappeared automatically). This function is obviously L-periodic, hence L-elliptic, and odd. By integration, we get

$$\wp(z + \omega) = \wp(z) + C_\omega \quad (C_\omega \text{ a constant}).$$

Since $\wp$ is even, $z = -\omega/2$ gives $C_\omega = 0$ (we are in characteristic $\neq 2$).

While we are at it, we give the Laurent expansion of $\wp$ and $\wp'$ at the origin. By termwise differentiation, we see that

$$\wp^{(2k)}(z) = (2k+1)!\sum (z - \omega)^{-2k-2} \quad ,$$

so that the Taylor coefficients of the even function $f(z) = \wp(z) - z^{-2}$ are the

$$f^{(2k)}(0)/(2k)! = (2k+1)\cdot G_{2k+2}$$

where we have defined $G_{2k} = G_{2k}(L) = \sum' \omega^{-2k}$ (for $k > 2$). We eventually get

(1.5) $$\wp(z) = z^{-2} + \sum_{k \geqslant 1} (2k+1) G_{2k+2}\, z^{2k} \qquad (0 < |z| < \operatorname*{Min}_{\omega \in L'} |\omega|),$$

and by derivation

(1.6) $$\wp'(z) = -2z^{-3} + \sum_{k \geqslant 1} (2k+1)2kG_{2k+2}\, z^{2k-1} \quad .$$

It is easy to give explicitely the divisors of several elliptic functions connected with $\wp$ or its derivative. For instance, $\wp'$ being odd, it follows that $\wp'$ vanishes at the points $a \in \mathbb{C}$ such that $-a \equiv a \bmod L$, i.e. $2a \in L$ (and $a \notin L$). A set of representatives of these points is $\omega_1/2$, $\omega_2/2$, $\omega_3/2$ with $\omega_3 = \omega_1 + \omega_2$. Since $\wp'$ has only a triple pole on $\mathbb{C}/L$, it can have only three zeros, and we must have

(1.7) $$\operatorname{div}(\wp') = -3(0) + (\tfrac{\omega_1}{2}) + (\tfrac{\omega_2}{2}) + (\tfrac{\omega_3}{2}) .$$

Similarly, the elliptic function $\wp - \wp(a)$ (for $a \notin L$) has only a double pole on $\mathbb{C}/L$, hence it has only the two zeros -a and a. At first, this is true for $a \not\equiv -a \bmod L$, but if $2a \in L$, $\wp'(a) = 0$ shows that the zero a of $\wp - \wp(a)$ is at least double, hence of order two, and

(1.8) $$\mathrm{div}(\wp - \wp(a)) = -2(0) + (a) + (-a) .$$

(The two symmetric zeros of $\wp$ are not explicitely known.)

We are now able to derive the explicit structure of the field of all elliptic functions.

(1.9) Theorem 2. The field of all L-elliptic functions is

$$\mathbb{C}(\wp,\wp') \cong \mathbb{C}(X)[Y]/(Y^2 - 4X^3 + g_2X + g_3) \quad ,$$

where $\wp = X$, $\wp'$ is identified with the image of Y in the quotient, and $g_2 = 60G_4 = 60\sum'\omega^{-4}$, $g_3 = 140G_6 = 140\sum'\omega^{-6}$ satisfy

$$\Delta = g_2^3 - 27g_3^2 \neq 0 \quad .$$

Moreover, the subfield of even elliptic functions is $\mathbb{C}(\wp)$.

Proof. Let f be any elliptic function. We can write f as sum of an even elliptic function and an odd elliptic function $f = f_1 + f_2$ (indeed put $f_1(z) = \frac{1}{2}(f(z) + f(-z))$ and $f_2 = f - f_1$). Since $\wp'$ is odd, we can write $f = f_1 + \wp'(f_2/\wp')$ with $f_2/\wp'$ even elliptic function. This already shows that the field of all elliptic functions is a quadratic extension of the subfield of even elliptic functions. We suppose now f is even. We let $\nu_i = \mathrm{ord}_{a_i}(f)$ if $a_i \not\equiv -a_i \pmod L$ and $\nu_i = \frac{1}{2}\mathrm{ord}_{a_i}(f)$ if $2a_i \in L$. These are integers since f being even has a pole or a zero of even order at the points ω_i . We consider then the product $g = \prod(\wp - \wp(a_i))^{\nu_i}$ extended over a set of representatives of the classes $\{a_i,-a_i\}$ with $a_i \notin L$. By construction, f and g have divisors with same coefficients, at least for all the points (a), where $a \notin L$, but the coefficient of (0) must also be the same in the two divisors by condition c) of Theorem 1. It follows that f is a constant multiple of g, hence a rational function in $\wp$. To prove the (quadratic) algebraic relation satisfied by $\wp$ and $\wp'$, we use the

Laurent expansions (1.5) and (1.6). Only the first few coefficients matter :

$$\wp(z) = z^{-2} + 3G_4 z^2 + 5G_6 z^4 + \mathcal{O}(z^6) \quad ,$$

$$\wp'(z) = -2z^{-3} + 6G_4 z + 20G_6 z^3 + \mathcal{O}(z^5) \quad ,$$

hence

$$\wp'(z)^2 = 4z^{-6} - 24G_4 z^{-2} - 80G_6 + \mathcal{O}(z^2) \ ,$$

$$\wp(z)^3 = z^{-6} + 9G_4 z^{-2} + 15G_6 + \mathcal{O}(z^2) \quad ,$$

Thus, $\wp'(z)^2 - 4\,\wp(z)^3 = -60G_4 z^{-2} - 140G_6 + \mathcal{O}(z^2) =$

$$= -g_2\,\wp(z) - g_3 + h(z) \quad ,$$

where h is L-elliptic, holomorphic, and $\mathcal{O}(z^2)$. This proves h = 0. To prove that the discriminant Δ does not vanish, we use a different form of the algebraic relation between $\wp$ and $\wp'$. Since we know the divisor of $\wp'$ by (1.7), we also know that of $\wp'^2$, hence we know all zeros of the cubic polynomial $4\wp^3 - g_2\wp - g_3$. They are the $e_i = \wp(\frac{\omega_i}{2})$ for i = 1,2,3, and are all double zeros. By comparing the leading terms, we get the equality

$$\wp'^2 = 4(\wp - e_1)(\wp - e_2)(\wp - e_3) \quad .$$

(Note that this is also a special case of the explicit expression of any even elliptic function as rational function of $\wp$ as derived above.)

By (1.8) we have

(1.10) $$\mathrm{div}(\wp - e_i) = -2(0) + 2(\tfrac{\omega_i}{2}) \qquad (\text{for } i = 1,2,3) \quad .$$

As the points $\omega_i/2$ are incongruent mod L , we conclude that these divisors are all distinct. In particular the functions $\wp - e_i$ must all be distinct. This proves that the three complex numbers e_i are distinct, and the discriminant $\Delta = \prod_{i \neq j}(e_i - e_j) \neq 0$. As is well known, this discriminant can be expressed in function of the coefficients of the equation as $g_2^3 - 27g_3^2$. Since the coefficient of $\wp^2$ in the cubic equation is 0, we also happen to see $e_1 + e_2 + e_3 = 0$, q.e.d.

Let $X_0 \subset \mathbb{C}^2$ be the curve of equation $y^2 = 4x^3 - g_2x - g_3$. I claim that it is non-singular (when $g_2^3 - 27g_3^2 \neq 0$). Indeed, the singular points would satisfy

$$0 = \frac{\partial}{\partial x}(y^2 - 4x^3 + g_2x + g_3) = -12x^2 + g_2 \quad ,$$

$$0 = \frac{\partial}{\partial y}(\ldots) = 2y \quad ,$$

hence would be on the x-axis with $x = \pm(g_2/12)^{\frac{1}{2}}$. But these points are not on the curve because $g_2^3 - 27g_3^2 \neq 0$, as a short computation shows.

We can define a mapping $\varphi_0 : \mathbb{C} - L \longrightarrow X_0 \subset \mathbb{C}^2$ by $\varphi_0(z) = (\wp(z), \wp'(z))$. It gives a bijection between E_0 and X_0 where $E_0 = E - \{0\}$ (and $E = \mathbb{C}/L$ as before). Indeed, $\wp$ assumes every complex value exactly twice with $\wp(z) = \wp(-z)$, and $\wp'$ separates z and -z (all this mod L). This holomorphic map has a holomorphic inverse $\varphi_0^{-1} : X_0 \longrightarrow E_0$ because its derivative $z \longmapsto (\wp'(z), \wp''(z))$ never vanishes (the only points z where $\wp'(z) = 0$ are the $\omega_i/2$ which are simple zeros of $\wp'$, hence $\wp''(\omega_i/2) \neq 0$).

To go from the affine curve X_0 to its natural "completion" (or "compactification") we embed $\mathbb{C}^2$ into the projective complex plane $\mathbb{P}^2(\mathbb{C}) = \mathbb{C}^3 - \{0\}/\text{homoth.}$, by $(x,y) \longmapsto (x,y,1) = (\lambda x, \lambda y, \lambda)$, and we define $X \subset \mathbb{P}^2(\mathbb{C})$ by the homogeneous equation $Y^2T = 4X^3 - g_2XT^2 - g_3T^3$ (I hope that there will be no confusion between the indeterminate X and the projective curve). Then we can extend φ_0 to φ as follows.

(1.11) Corollary. *Let* $\varphi : z \longmapsto (z^3\wp(z), z^3\wp'(z), z^3)$ *be the mapping*

$$\mathbb{C} \longrightarrow X \subset \mathbb{P}^2(\mathbb{C})$$

extending φ_0. *Then* φ *gives an analytic isomorphism* $E \xrightarrow{\sim} X$, *and this last curve is non-singular.*

All assertions have been proved, or can be checked directly. We only note that the point at infinity on X is $(0,1,0) = (0,\lambda,0)$, and that it is non-singular (we shall see later that it is an ordinary flex on X).

A small point deserves mention. Since $\mathbb{C}$ is simply connected, and φ is a local isomorphism, we can identify φ with the universal covering map for X. This universal covering map is known explicitely by means of the $\wp$ and $\wp'$-functions. This is an explicit uniformization for the curve $y^2t = 4x^3 - g_2xt^2 - g_3t^3$. (We shall see later, that for any couple (g_2, g_3) of complex values such that $g_2^3 - 27g_3^2 \neq 0$, there exists a lattice $L \subset \mathbb{C}$ such that $g_2 = g_2(L)$ and $g_3 = g_3(L)$, so that the uniformization can be made explicitely for all those cubics. If, more generally, X is a non-singular projective curve of genus $\geqslant 2$, the theorem of uniformization of Poincaré-Koebe asserts that the universal covering of X is the open unit disc in $\mathbb{C}$, hence X is isomorphic to the quotient of the disc by a discrete subgroup of automorphisms of the disc. But the link between X - or its defining equation - and this discrete subgroup cannot be given explicitely (yet), except in very special cases.)

We can transform slightly the equation $y^2 = 4x^3 - g_2x - g_3$ of X_o. First, we replace y by y/2 and define new constants $A = -g_2/4$ and $B = -g_3/4$, so that we get the new equation

(1.12) $$y^2 = x^3 + Ax + B \quad \text{with} \quad 4A^3 + 27B^2 \neq 0 .$$

Moreover, we know the roots of the right-hand side, they are the e_i (see proof of Th.2), so that this equation (1.12) can be rewritten

(1.12)' $$y^2 = (x-e_1)(x-e_2)(x-e_3) \quad \text{with} \quad \prod_{i \neq j} (e_i - e_j) \neq 0 .$$

It is still useful to make a further change of variables by putting $\xi = (x-e_2)(e_1-e_2)^{-1}$, $\eta = (e_1-e_2)^{-3/2}y$, and $\lambda = (e_3-e_2)(e_1-e_2)^{-1}$, whence we get the new equation

(1.13) $$\eta^2 = \xi(\xi - 1)(\xi - \lambda) \quad \text{with} \quad \lambda(\lambda - 1) \neq 0 .$$

This is the reduced form of the cubic given by Legendre. This equation has only one parameter, namely λ (but different values of λ can give the "same" - isomorphic - curves).

By using the isomorphism between E and the curve X, we consider X as an abelian group.

(1.14) Theorem 3 (Addition). a) Geometric form: three points on X have sum zero whenever they are colinear (they lie on a line).

b) Analytic form: Let $u \neq \pm v$ (mod L) be two complex numbers. Then

$$\wp(u+v) = -\wp(u) - \wp(v) + \frac{1}{4}\left(\frac{\wp'(u) - \wp'(v)}{\wp(u) - \wp(v)}\right)^2 .$$

A picture will help. We suppose that the three roots of the cubic polynomial $4x^3 - g_2x - g_3$ are real. Then it is possible to choose a real hyperplane in $\mathbb{C}^2$ containing the curve X_o . The complex conjugation induces a symmetry with respect to a plane which we take horizontal :

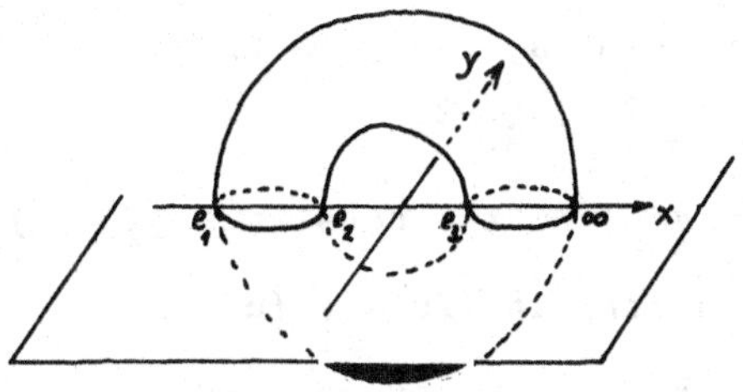

complex curve X

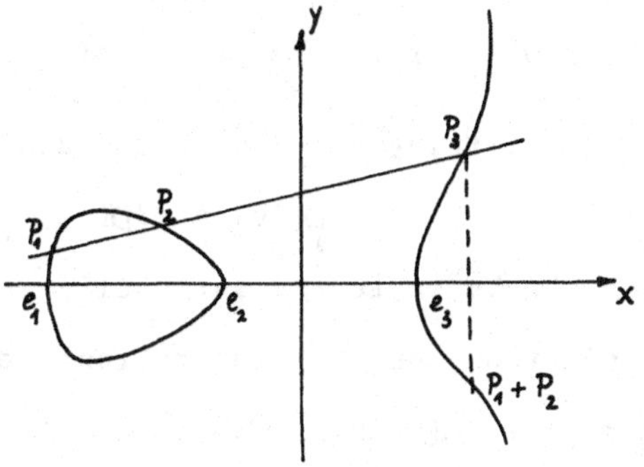

real points of X_o

Proof. a) There is no loss in generality in supposing that the first two points have distinct first coordinates: $P_1 = (x_1,y_1)$, $P_2 = (x_2,y_2)$ with $x_1 \neq x_2$. Let $y = mx + h$ be the equation of the straight line through these two points. The elliptic function $\wp' - m\wp - h$ has a pole of order three (at the origin of E), hence has three zeros, two of which u, v are known, namely those u and v for which $\varphi(u) = P_1$ and $\varphi(v) = P_2$. We call $P_3 = \varphi(w)$ where w is the third zero of that elliptic function. The last property (Th.1, d) of divisors of elliptic functions shows that $u + v + w = 0$ (mod L), hence the sum of the P_i's is zero. Conversely, if three points are on X and on a line, they must be the three zeros of an elliptic function (given by the equation of the line as above), hence have sum zero.

b) We keep the notations of above, but add $P_1 = (\wp(u), \wp'(u))$, and $P_2 = (\wp(v), \wp'(v))$, so that

$$m = \frac{y_2 - y_1}{x_2 - x_1} = \frac{\wp'(u) - \wp'(v)}{\wp(u) - \wp(v)} \quad .$$

Also we can write for the coordinates of the intersection points of X_o and the line

$$\begin{aligned} y^2 - (mx + h)^2 &= 4x^3 - g_2x - g_3 - m^2x^2 - 2mhx - h^2 = \\ &= 4(x^3 - x^2m/4 + \ldots) = \\ &= 4(x - \wp(u))(x - \wp(v))(x - \wp(w)). \end{aligned}$$

This gives $\wp(u) + \wp(v) + \wp(w) = m^2/4$ hence the announced formula by the formula for m, the parity of $\wp$ and $w = -u-v \pmod{L}$. q.e.d.

By letting $v \longrightarrow u$ in the addition formula, one gets also

(1.15) $$\wp(2u) = -2\wp(u) + \frac{1}{4}\wp''(u)^2 \wp'(u)^{-2}$$

which can be expressed rationally in $\wp$, $\wp'$ (and the constants g_2, g_3). By induction, one can get rational expressions in $\wp(u)$, $\wp'(u)$, g_2 and g_3 for $\wp(nu)$ (and also for $\wp'(nu)$) where n is any positive integer.

The function $\wp$ of Weierstrass having zero residues at all its poles, has a well-defined primitive over the simply connected space $\mathbb{C}$. We can find one by integrating term by term the definition of $\wp$ as infinite sum, adjusting the constants of integration of all terms so as to get a convergent series. Thus we define

(1.16) $$\xi(z) = -z^{-1} - \sum{}'((z - \omega)^{-1} + \omega^{-1} + z\omega^{-2}) \quad .$$

(Classically, this primitive is $-\zeta(z)$ with the notations of Weierstrass but we refrain from using this zeta notation, because certain formulas use also Riemann's zeta function, hence create difficulties.)

The integration can be carried one step further, but the non-zero residues of ξ make its primitive logarithmically ramified at the points of the lattice, so we define this primitive to be log(f(z)), with a certain function f, easily defined by a canonical Weierstrass' product. It is again a habit to define $1/f = \sigma$ by

(1.17) $$\sigma(z) = z \prod{}'(1 - z/\omega)\exp\left(\frac{z}{\omega} + \frac{1}{2}\left(\frac{z}{\omega}\right)^2\right) .$$

By definition, the functions ξ and σ are odd functions. σ is holomorphic with simple zeroes at the points of the lattice (it is an entire function). Also by definition

(1.18) $$\xi'(z) = \wp(z) \text{ and } (\sigma'(z)/\sigma(z))' = -\wp(z) .$$

The periodicity of $\wp$ gives by integration $\xi(z) = \xi(z+\omega) + \eta(\omega)$. If we choose a basis ω_1, ω_2 of L as before, we put

(1.19) $$\xi(z) = \xi(z + \omega_i) + \eta_i \quad , \quad (\eta_i = -2\xi(\omega_i/2)) .$$

The two constants η_i are not arbitrary. An integration of $\xi(z)dz$ on the sides of a period parallelogram P_a gives

(1.20) $$\eta_1\omega_2 - \eta_2\omega_1 = 2\pi i .$$

This is a relation discovered by Legendre.

Moreover, the integration of (1.19) gives $\sigma(z+\omega_i) = C_i\sigma(z)e^{\eta_i z}$ with two constants C_i which can be computed by making $z = -\omega_i/2$ and using the fact that σ is odd (also note that $\sigma(\omega_i/2) \neq 0$ because the only zeros of σ are the points of the lattice). We find

(1.21) $$\sigma(z + \omega_i) = -e^{\eta_i(z+\frac{1}{2}\omega_i)}\sigma(z) .$$

The function σ is especially useful to study divisors on E. Indeed, let $\underline{d} = \sum d_k(a_k) \in \mathrm{Div}(E)$ be any divisor, and take representatives $a_k \in \mathbb{C}$ of the points of E appearing in $\underline{d}$. Then define

$$f_{\underline{d}} = \prod \sigma(z - a_k)^{d_k} .$$

This is a meromorphic function in $\mathbb{C}$ which depends not only on $\underline{d}$ but also on the selected representatives a_k of the points appearing in $\underline{d}$, a fact which we keep in mind, but not in notation! By (1.21) we have

$$f_{\underline{d}}(z+\omega_i)/f_{\underline{d}}(z) = \prod (-1)^{d_k} \cdot e^{d_k \eta_i (z - a_k + \frac{1}{2}\omega_i)}$$

$$= (-1)^{\Sigma d_k} \cdot e^{\eta_i (z + \frac{1}{2}\omega_i)\Sigma d_k} \cdot e^{-\eta_i \Sigma d_k a_k} .$$

In this formula, we see that the two conditions

$$\deg(\underline{d}) = \sum d_k = 0 , \quad \sum d_k a_k = 0 ,$$

imply that $f_{\underline{d}}$ is an <u>elliptic</u> function. It is also clear that $\mathrm{div}(f_{\underline{d}}) = \underline{d}$ because σ has simple poles at the points of the lattice L only. Finally, we observe that if $\underline{d}$ is any divisor satisfying the two conditions c),d) of Th.1, say

$$\underline{d} = \sum d_k(a_k) , \quad \sum d_k = 0 , \quad \sum d_k a_k = \omega \in L , \quad (k \geq 1),$$

then we can define representatives

$$a_o = a_1 - \omega , \quad d'_k = 1 \quad \text{for } k = 0 ,$$

$$d'_k = d_k - 1 \text{ for } k = 1 ,$$

$$d'_k = d_k \text{ for } k \geq 2 ,$$

satisfying $\sum d'_k a_k = 0$ and $\underline{d} = \sum_{k\geq 1} d_k(a_k) = \sum_{k \geq 0} d'_k(a_k)$, so that we can construct an elliptic function $f_{\underline{d}}$ with divisor $\underline{d}$.

(1.22) <u>Theorem 4</u>. <u>Let</u> $\underline{d} = \sum d_k(a_k) \in \mathrm{Div}(E)$ <u>be a divisor on</u> E. <u>There exists an elliptic function with divisor</u> $\underline{d}$ <u>(unique up to a non-zero multiplicative constant) if and only if the two following conditions are satisfied</u> : $\deg(\underline{d}) = 0$ and $\sum d_k a_k \in L$. <u>If these conditions are satisfied, it is possible to choose representatives such that</u> $\sum d_k a_k = 0$, <u>and for this choice, this elliptic function is proportional to</u> :

$$f_{\underline{d}}(z) = \prod \sigma(z - a_k)^{d_k} .$$

As application, we quote a formula. We know explicitely the divisor of $\wp - \wp(a)$, hence we get

$$\wp(z) - \wp(a) = C\sigma(z-a)\sigma(z+a)\sigma(z)^{-2} .$$

The constant C can be determined by looking at the asymptotic behavior when $z \longrightarrow 0$ (σ is odd and $a \notin L$ implies $\sigma(a) \neq 0$) whence

$C = -\sigma(a)^{-2}$. We have proved

(1.23) $$\wp(u) - \wp(v) = -\sigma(u-v)\sigma(u+v)\sigma(u)^{-2}\sigma(v)^{-2} ,$$

for $u,v \neq 0 (\mathrm{mod}\ L)$. Similarly,

(1.23)' $$\wp'(z) = 2\sigma(z)^{-3}\prod_{i=1}^{3}\sigma(z-\tfrac{1}{2}\omega_i)\sigma(\tfrac{1}{2}\omega_i)^{-1} , \qquad (z \notin L),$$

and this is also

$$= -(\sigma'(z)/\sigma(z))'' \quad \text{by (1.18) !}$$

The Theorem 4 has an important Corollary which gives the structure of the abelian group of classes of divisors mod the divisors of elliptic functions. We define P(E) to be the subgroup of Div(E) consisting of divisors of elliptic functions, also called <u>principal divisors</u>, or divisors linearly equivalent to zero. This is a subgroup of the group $\mathrm{Div}^{o}(E)$ of divisors of degree zero. We have just seen, that in $\mathrm{Div}^{o}(E)$, P(E) is characterised by $\sum d_k a_k = 0 \in E$ for $\underline{d} = \sum d_k(a_k)$. The Picard group Pic(E) of E is by definition the quotient group

$$\mathrm{Pic}(E) = \mathrm{Div}^{o}(E)/P(E)$$

(endowed with an analytic structure, as we shall show presently). (In the scheme terminology of Grothendieck, it would rather be the bigger quotient Div(E)/P(E), but we stick to the classical definition.) Then we define a mapping

$$\Phi : E \longrightarrow \mathrm{Div}^{o}(E)/P(E)$$
$$a \longmapsto (a) - (0) \quad \mathrm{mod}\ P(E) \quad .$$

This is a group homomorphism. Indeed

$$(a) - (0) + (b) - (0) - (a+b) + (0) =$$
$$= (a) + (b) - (a+b) - (0)$$

is a principal divisor by the criterium given by Th.4. Then we can see that Φ is bijective: $\Phi(a) = (a) - (0)$ is a principal divisor only when $a = 0$ (in E), so that Φ is injective, and moreover, if $\underline{d} = \sum d_k(a_k)$ has degree zero, it can be written

$$\underline{d} = \sum d_k((a_k) - (0)) = \sum d_k\Phi(a_k) = \Phi(\sum d_k a_k) ,$$

which proves Φ surjective.

(1.24) Corollary. The mapping $\Phi: E \longrightarrow \mathrm{Pic}(E) = \mathrm{Div}^{0}(E)/P(E)$

$a \longmapsto (a) - (0) \bmod P(E)$,

is a group isomorphism. In particular, Pic(E) is canonically an analytic group.

We mention a few exact sequences which connect now Div(E)/P(E) with the Pic(E). First we have the exact sequence of definition of Pic(E):

$$0 \longrightarrow P(E) \longrightarrow \mathrm{Div}^{0}(E) \longrightarrow \mathrm{Pic}(E) \longrightarrow 0 .$$

On the other hand, we have also an obvious exact sequence

$$0 \longrightarrow \underset{\underset{E}{\wr\wr}}{\mathrm{Pic}(E)} \longrightarrow \mathrm{Div}(E)/P(E) \xrightarrow{\deg} \mathbb{Z} \longrightarrow 0$$

so that Div(E)/P(E) is an extension of $\mathbb{Z}$ by a group isomorphic to E. We can bunch together the above exact sequences in a longer exact sequence

$$0 \longrightarrow P(E) \longrightarrow \mathrm{Div}^{0}(E) \longrightarrow \mathrm{Div}(E)/P(E) \xrightarrow{\deg} \mathbb{Z} \longrightarrow 0 .$$

The cohomological interpretation of these sequences will only be indicated in next section (on theta functions).

Let us observe that the homomorphism deg : $\mathrm{Div}(E) \longrightarrow \mathbb{Z}$, is nothing but the augmentation homomorphism $\mathbb{Z}[E] \longrightarrow \mathbb{Z}$, and on the other hand, the homomorphism $\sum d_k(a_k) \longmapsto \sum d_k a_k \in E$ is the canonical homomorphism $\mathbb{Z}[E] \longrightarrow E$ given by the (nonetheless canonical) $\mathbb{Z}$-module structure of the abelian group E. The principal divisors are the elements of the intersection of the two kernels of these homomorphisms

$$P(E) \longrightarrow \mathbb{Z}[E] \rightrightarrows \begin{matrix} \mathbb{Z} \\ E \end{matrix} .$$

This shows that an analogue of P(E) can be defined for any abelian group G.

An additive formula corresponding to the multiplicative (1.22) (Th.4) is due to Hermite. It can be explained quickly as follows.

Let f be an elliptic function with poles at points a_k with principal part

$$A_1(z - a)^{-1} + A_2(z - a)^{-2} + \dots + A_n(z - a)^{-n}$$

(where $a = a_k$ and $n = n(k)$). We define the meromorphic function

$$P_a(z - a) = -A_1\zeta(z - a) + A_2\wp(z - a) + \dots$$

$$\dots + (-1)^n A_n \wp^{(n-2)}(z - a)/(n-1)! \quad .$$

Then we add all these explicit functions (corresponding to the principal parts of f at the poles $a = a_k$)

$$g(z) = \sum_{a=a_k} P_a(z - a) \quad .$$

By construction f - g has no pole, and is an elliptic function, because the sum of the residues of f is zero. The announced formula follows

(1.25) $$f(z) = \sum_{a=a_k} P_a(z - a) + C \quad .$$

(1.26) <u>Addendum</u>. We look at $\wp$ as function on $E = \mathbb{C}/L$. It is a holomorphic mapping $\wp : E \longrightarrow \mathbb{C}\cup\{\infty\} = \underline{S}^2$ (the complex sphere of Riemann). We claim that it is a double sheeted "covering" of $\underline{S}^2$ with four (distinct) ramification points of order two $e_1, e_2, e_3, \infty \in \underline{S}^2$. Indeed, the valence of $\wp$ is two, and $\wp$ takes the values e_i with multiplicity two because $\wp'(\omega_i/2) = 0$ (the case ∞ is more obvious). This mapping shows that E can be constructed by glueing two Riemann spheres with two cuts (interchanging the sheets), one joining e_1 to e_2 and the other one joining e_3 to ∞. On the algebraic curve X, this mapping is explained as follows. It coïncides with the first coordinate mapping (on X_o)

$$X \longrightarrow \mathbb{P}^1(\mathbb{C}) = \underline{S}^2$$
$$(x,y,t) \longmapsto (x,t) \longmapsto x/t \quad .$$

(On X_o , we can take t = 1.) A natural generalization of this situation is to consider a projective curve X of homogeneous equation

$$y^2t^{2g-1} = \prod(x - e_it) \qquad (1 \leqslant i \leqslant 2g+1)$$

with the non-singularity condition for $t \neq 0$

$$\prod_{i \neq j} (e_i - e_j) \neq 0 \quad .$$

The mapping $X \longrightarrow \underline{S}^2$ defined again by $(x,y,t) \longmapsto (x,t) \longmapsto x/t$, is still a double sheeted "covering" with the $2g+2$ ramification points $e_1, \dots, e_{2g+1}, e_{2g+2} = \infty$ .The involutive automorphism of interchanging the sheets is $(x,y,t) \longmapsto (x,-y,t)$. This is an example of what is called a hyperelliptic curve.

(1.27) Open Problem. We have mentioned after (1.11) (p.I.10), the general problem of explicit uniformization of algebraic curves of genus $g \geq 2$. There is a still more general (unsolved!) problem for an explicit uniformization of curves with a given signature (we shall not define this notion) which gives an interesting problem for elliptic curves. In its simplest form, it can be stated as follows. Let E_0 be a punctured elliptic curve, i.e. an elliptic curve X minus one point, which we take as the origin. The fundamental group of X_0 is isomorphic to a free (non-abelian) group over two generators, and the universal covering (as analytic and topological space) of X_0 is known to be isomorphic to the upper complex line $H = \{z \in \mathbb{C} : \mathrm{Im}(z) > 0\}$ of Poincaré (or equivalently, the open unit disc in $\mathbb{C}$).How can one find explicitely a uniformizing map $\varphi : H \longrightarrow X_0$? In particular, X_0 is isomorphic to a quotient of H by a discrete subgroup of automorphisms Γ (with finite volume quotient : Γ is a Fuchsian group of first kind, isomorphic to $\pi_1(X_0) \cong \mathbb{Z} * \mathbb{Z}$, free amalgamation). How can one determine Γ in $SL(2,\mathbb{R})$ by means of the equation of X_0 ?

(1.28) Erratum. p.I.8, bottom. If we define $D = \prod_{i \neq j} (e_i - e_j)$ and $\Delta = g_2^3 - 27g_3^2$, then Δ is proportional to D with a coefficient independent of L : $\Delta = -16 \cdot D$.

2. Theta functions (Jacobi)

Several proportional bilinear alternate forms appear in the theory, e.g. when one generalizes the relation (1.20) (Legendre's relation). So we start by showing that a lattice $L \subset \mathbb{C}$ determines a canonical one

$$B : \mathbb{C} \times \mathbb{C} \longrightarrow \mathbb{R} \qquad \mathbb{R}\text{-bilinear} ,$$

as follows. Let ω_1, ω_2 be a basis of L. By interchanging ω_1 and ω_2, we may suppose $\mathrm{Im}(\omega_2/\omega_1) > 0$ (we could also change the sign of ω_1). Such a basis is said to be <u>direct</u> or <u>positively oriented</u>. Then for any $x = x_1\omega_1 + x_2\omega_2$ and $y = y_1\omega_1 + y_2\omega_2$ in $\mathbb{C}$, we define

$$(2.1) \qquad B(x,y) = B_L(x,y) = \det\begin{pmatrix} x_1 & x_2 \\ y_1 & y_2 \end{pmatrix} .$$

By definition, it is obvious that $B(\omega_1,\omega_2) = 1$. More generally,

$$B(\omega_1',\omega_2') = \pm 1 \text{ (for } \omega_i' \in L) \text{ whenever } (\omega_i') \text{ is a basis of } L ,$$

and more precisely,

$$B(\omega_1',\omega_2') = 1 \text{ (ibid.) whenever } (\omega_i') \text{ is a direct basis of } L .$$

This shows that B is defined independently from the choice of the direct basis ω_1,ω_2. Thus B is an $\mathbb{R}$-bilinear alternate form on $\mathbb{C}$. But $(x,y) \longmapsto \mathrm{Im}(\bar{x}y) = (\bar{x}y - x\bar{y})/2i$ is also such a form. Hence B must be proportional to this form, and exactly

$$B(x,y) = \mathrm{Im}(\bar{x}y)/\mathrm{Im}(\bar{\omega}_1\omega_2) ,$$

for any direct basis ω_1,ω_2. We can express slightly differently this constant of proportionality. Indeed,

$$\mathrm{Im}(\bar{\omega}_1\omega_2) = \mathrm{Im}(\omega_1\bar{\omega}_1\omega_2/\omega_1) = |\omega_1|^2\mathrm{Im}(\omega_2/\omega_1) = |\omega_1|^2\mathrm{Area}(1,\omega_2/\omega_1) =$$

$$= \mathrm{Area}(\omega_1,\omega_2) = S > 0 .$$

We define this constant S to be the <u>mesh</u> of the lattice L. Hence

$$(2.2) \qquad B(x,y) = \frac{1}{2iS}(\bar{x}y - x\bar{y}) = \frac{1}{S}\mathrm{Im}(\bar{x}y) .$$

Let us look at the bilinear form $(x,y) \longmapsto B(x,iy)$. From (2.2), we see that $B(x,iy) = S^{-1}\mathrm{Im}(i\bar{x}y) = S^{-1}\mathrm{Re}(\bar{x}y)$ is symmetric, $\mathbb{R}$-bilinear and positive definite : $B(x,ix) > 0$ for $0 \neq x \in \mathbb{C}$. To sum up, the $\mathbb{R}$-bilinear form B has the following properties

(2.3) a) $B(y,x) = -B(x,y)$ (B is alternate) ,

b) $B(L,L) \subset \mathbb{Z}$ (B is integral on L) ,

c) $(x,y) \mapsto B(x,iy)$ is symmetric positive definite.

(2.4) Remark. When L is a lattice in $\mathbb{C}^n$, the existence of an $\mathbb{R}$-bilinear form $B : \mathbb{C}^n \times \mathbb{C}^n \longrightarrow \mathbb{R}$ satisfying the three conditions of (2.3) gives a criterium for the existence of sufficiently (global) meromorphic functions on $\mathbb{C}^n/L$ to be able to separate the points. When this is the case, $\mathbb{C}^n/L$ has a projective embedding and is called an abelian variety (over $\mathbb{C}$). Any form B satisfying (2.3) is called Riemann form on $\mathbb{C}^n$. We have just seen that for n = 1, this condition is always satisfied (more precisely : canonically satisfied).

Now we start with the subject of this section. It is impossible to find holomorphic (non-constant) L-elliptic functions, but we can try to find holomorphic functions which satisfy equations of the form

$$\theta(z + \omega) = e^{A(\omega)z + B(\omega)}\theta(z) \qquad (\omega \in L).$$

The quotient of two such holomorphic functions with same A and B will obviously give L-elliptic functions. The σ function of Weierstrass has the above property, and the transformation formula (1.21) leads to another definition of the constants (also the relation of Legendre (1.20) shows that it might be useful to introduce a π in η to get a more algebraic form of this relation). We adopt the following

(2.5) Definition. A theta function θ of type (h,a) with respect to L is a meromorphic function on the complex line $\mathbb{C}$ satisfying

$$\theta(z + \omega) = a(\omega)e^{\pi h(\omega)(z+\frac{1}{2}\omega)}\theta(z) \qquad (\omega \in L) .$$

We note that a theta function θ is characterized by the fact that $(\theta'/\theta)'$ is an elliptic function. We shall later on be mainly interested in holomorphic theta functions, but it would be too much of a strain to restrict ourselves to them from the beginning. It follows immediately from the definition that if we have theta functions θ of type (h,a) and $\tilde{\theta} \neq 0$ of type $(\tilde{h},\tilde{a})$, then

$$\theta\tilde{\theta} \text{ is a theta function of type } (h + \tilde{h}, a\tilde{a}) \quad ,$$
$$\tilde{\theta}^{-1} \quad " \quad " \quad " \quad " \quad " \quad (-\tilde{h},\tilde{a}^{-1}) \quad ,$$
$$\theta/\tilde{\theta} \quad " \quad " \quad " \quad " \quad " \quad (h - \tilde{h}, a/\tilde{a}) \ .$$

Examples of theta functions abound. First comes to mind the σ function of Weierstrass, and all the products of its translates. In particular all elliptic functions are theta functions. Moreover, the exponentials $z \longmapsto e^{\alpha z^2 + \beta z + \gamma}$ are also theta functions, as is easily checked. Because the multiplicator $a(\omega)\exp(\pi h(\omega)(z + \frac{1}{2}\omega))$ is an entire function with no zero, we can define unambiguously <u>the divisor of a theta function</u> $\mathrm{div}(\theta)$ as element of $\mathrm{Div}(E)$ (the order of θ at a point a is the same as the orders of θ at all translates $a + \omega$ of a by all periods $\omega \in L$). For example

$$\mathrm{div}(\sigma) = (0) \quad \text{(one point of E with multiplicity one)} \ ,$$
$$\mathrm{div}(z \longmapsto e^{\alpha z^2 + \beta z + \gamma}) = 0 \ \text{(neutral element of Div(E))}.$$

For this reason, the exponentials of quadratic polynomials are called <u>trivial</u> theta functions.

The type of a theta function θ is a couple of mappings

$$h : L \longrightarrow \mathbb{C} \quad , \quad a : L \longrightarrow \mathbb{C}^{\times} \ ,$$

with interesting properties. First, replacing ω by $\omega + \omega'$ in (2.5) shows that h is a homomorphism, hence $\mathbb{Z}$-linear, hence can be uniquely extended in $\mathbb{R}$-linear homomorphism $\mathbb{C} \longrightarrow \mathbb{C}$ still denoted h. Its <u>coboundary</u> ∂h is defined by $\partial h(x,y) = h(x)y - xh(y)$, so that ∂h is $\mathbb{R}$-bilinear alternate $\mathbb{C} \times \mathbb{C} \longrightarrow \mathbb{C}$. It must be proportional to the canonical B defined in (2.2), the constant being determined by the value $\partial h(\omega_1,\omega_2)$ for a direct basis (ω_i) of L. But

$$\pi(h(\omega_1)\omega_2 - \omega_1 h(\omega_2)) = \oint_{\partial P_a} \theta'(z)/\theta(z)dz =$$

$$= 2\pi i \deg(\mathrm{div}\theta) \quad .$$

Hence we have proved

(2.6) $\quad \partial h(x,y) = 2i\cdot\deg(\mathrm{div}\theta)\cdot B(x,y) = S^{-1}\deg(\mathrm{div}\theta)\cdot(\bar{x}y - x\bar{y}) \quad .$

This is a generalization of Legendre's relation (1.20).

(2.7) Definition. A theta function θ of type (h,a) is said to be reduced when $\theta \neq 0$ and

$$|a(\omega)| = 1 \text{ and } h(\omega)\omega \text{ is real (for any } \omega \in L).$$

Every theta function $\theta \neq 0$ can be uniquely written in the form

$\theta = \theta_o \cdot \theta_{red}$ where θ_o is a trivial theta function such that $\theta_o(0) = 1$ (hence of the form $\theta_o(z) = \exp(\alpha z^2 + \beta z)$) and θ_{red} a reduced theta function. It is indeed sufficient to adjust the two constants α and β suitably. In particular $\mathrm{div}(\theta) = \mathrm{div}(\theta_{red})$.

(2.8) Proposition. Let θ be a reduced theta function of type (h,a) and put $d = \deg(\mathrm{div}\theta)$. Then $h(x) = dS^{-1}\bar{x}$. If moreover $\tilde{\theta}$ is another reduced theta function,

a) $\theta\tilde{\theta}$ and $\theta/\tilde{\theta}$ are reduced theta functions ,

b) $\mathrm{div}(\theta) = \mathrm{div}(\tilde{\theta})$ implies $\theta = c\tilde{\theta}$ $(c \in \mathbb{C})$.

Proof. We can write the $\mathbb{R}$-linear extension of h in the form $h(z) = A\bar{z} + Bz$ (with constants A and B). Because $h(z)z = A|z|^2 + Bz^2$ must be real for all z (as follows immediately from $h(\omega)\omega$ real for all $\omega \in L$ and real linearity), we see that $B = 0$ and A is real. Then (2.6) gives $h(z) = dS^{-1}\bar{z}$. Then a) is trivial, and to prove b) we use it, and replace θ by $\theta/\tilde{\theta}$. We have to prove

$$\mathrm{div}(\theta) = 0 \text{ and } \theta \text{ reduced implies } \theta = \text{constant}.$$

By the first part of the proposition, $\mathrm{div}(\theta) = 0$ implies $d = 0$, hence $h = 0$. Thus $\theta(z + \omega) = a(\omega)\theta(z)$ and because $|a| = 1$, we see that θ is an entire bounded function ($|\theta|$ is periodic). Liouville's theorem shows then that θ must be constant.

Let now Θ_{red} be the set (a group) of reduced theta functions. Then

$$\mathbb{P}(\Theta_{red}) = \Theta_{red} \text{ mod. scalars} = \Theta_{red}/\mathbb{C}^{\times}$$

is a group (if $\theta \neq 0$ is reduced, so is θ^{-1}).

(2.9) Corollary. The canonical mapping

$$\text{div} : \mathbb{P}(\Theta_{red}) \longrightarrow \text{Div}(E)$$

$$\theta \text{ mod homoth.} \longmapsto \text{div}\theta \quad ,$$

is a group isomorphism.

Proof. This map is a group homomorphism because $\text{div}(\theta\tilde{\theta}) = \text{div}\theta + \text{div}\tilde{\theta}$. The map is surjective by using the reduced theta function σ_{red} corresponding to σ and products of translates $\prod \sigma_{red}(z - a_k)^{d_k}$. The map is injective by b) of Prop. above.

It is thus equivalent to work with reduced non-zero theta functions (mod multiplicative constants) or with divisors on E. The proof of (1.22) only used one facet of this general fact.

(2.10) Theorem (Appel-Goursat). Let $\underline{d}$ be a divisor on E of degree $d \geq 1$, and choose a theta function θ with $\text{div}(\theta) = \underline{d}$. Then the complex vector space of holomorphic theta functions having same type as θ is of dimension $\deg(\underline{d}) = d$.

Proof. Let (h,a) be the type of θ, and define

$$T(u) = a(\omega_1)^{-u}\exp(-\tfrac{1}{2}\pi h(\omega_1)\omega_1 u^2)\cdot\phi(\omega_1 u) \quad .$$

for every holomorphic theta function ϕ having type (h,a). We shall show that such functions ϕ exist! After a small computation, one finds

$$(*) \qquad T(u + 1) = T(u) \quad .$$

Also, replacing u by $u + \tau$ with $\tau = \omega_2/\omega_1$ (where we suppose the basis (ω_i) of L direct, hence $\text{Im}(\tau) > 0$), we see that

$$T(u + \tau) = T(u)a(\omega_2)a(\omega_1)^{-\tau} \exp\pi u(h(\omega_2)\omega_1 - h(\omega_1)\omega_2) \cdot$$

$$\cdot\exp(\tfrac{1}{2}\pi(h(\omega_2)\omega_2 - h(\omega_1)\omega_1\tau^2)) \quad ,$$

that is,

$$T(u + \tau) = T(u)a(\omega_2)a(\omega_1)^{-\tau}\exp(-\pi u\partial h(\omega_1,\omega_2) - \tfrac{1}{2}\pi\tau\partial h(\omega_1,\omega_2)) \quad .$$

We define $C = a(\omega_2)/a(\omega_1)^{\tau}$ and $q = \exp(\pi i\tau)$, so that the above transformation formula is more simply

$$(**) \qquad T(u + \tau) = Cq^{-d}\exp(-2\pi idu)T(u) \quad .$$

If ϕ, hence T, exist, we see from (*) that there will exist a holomorphic function F in $\mathbb{C} - \{0\}$ such that

$$T(u) = F(\zeta) = F(e^{2\pi iu}).$$

The Laurent expansion of F at the isolated singularity $0 \in \mathbb{C}$ will have the form

$$F(\zeta) = \sum_{-\infty}^{\infty} a_\ell \zeta^\ell \quad .$$

But (**) implies certain relations for the coefficients a_ℓ. Indeed $\exp(2\pi i(u+\tau)) = e^{2\pi iu}e^{2\pi i\tau} = q^2\zeta$ and $\exp(-2\pi idu) = \zeta^{-d}$ gives

$$\sum_{-\infty}^{\infty} a_\ell q^{2\ell}\zeta^\ell = \sum_{-\infty}^{\infty} a_\ell Cq^{-d}\zeta^{-d}\zeta^\ell \quad .$$

This in turn gives, by comparing the coefficients of ζ^ℓ,

$$a_\ell q^{2\ell} = a_{\ell+d}Cq^{-d} \qquad \text{or}$$

$$a_{\ell+d} = C^{-1}q^{2\ell+d}a_\ell \quad .$$

Hence

$$a_{\ell+dk} = a_\ell C^{-k}q^{2\ell+d}\cdot q^{2(\ell+d)+d}\cdots q^{2[\ell+(k-1)d]+d} \quad .$$

But

$$\underbrace{2\ell+d} + \underbrace{2\ell+3d} + \ldots + \underbrace{2\ell+(2k-1)d} = 2\ell k + d(1 + 3 + \ldots + (2k-1)) =$$

$$= 2\ell k + dk^2 \quad .$$

Finally, we see that

$$a_{\ell+kd} = a_\ell C^{-k}q^{2\ell k}q^{dk^2} \quad ,$$

and by summing these coefficients together, we get

$$F(\zeta) = \sum_{-\infty}^{\infty} a_\ell\zeta^\ell = \sum_{\ell=0}^{d-1} a_\ell\zeta^\ell \sum_{k\in\mathbb{Z}} C^{-k}q^{2\ell k+dk^2}\zeta^{kd} \quad ,$$

$$F(\zeta) = \sum_{\ell=0}^{d-1} a_\ell F_\ell(\zeta)\zeta^\ell \quad ,$$

with

$$F_\ell(\zeta) = \sum_{k\in\mathbb{Z}} C^{-k} q^{dk^2+2\ell k} \zeta^{dk} \quad .$$

Now all these series converge absolutely because

$$\mathrm{Im}(\tau) > 0 \text{ implies } |q| = |e^{\pi i\tau}| = e^{-\pi \mathrm{Im}(\tau)} < 1$$

and the term $|q^{dk^2}| = |q|^{dk^2}$ decreases faster than all the others for $|k| \to \infty$. This shows that all $F_\ell(\zeta)$ are well defined in $\mathbb{C}^\times$ for $\ell = 0, 1, \ldots, d-1$. We could have started by these functions, and in reverse direction define T and finally ϕ. As the space of the F's is d dimensional, so must be the space of the T, and also the space of the ϕ, which proves the theorem.

It is interesting to look at the case of $\theta = \sigma$, so that $\underline{d} = (0)$ has degree precisely d = 1. With the notations of the above proof,

$$F_0(\zeta) = \sum_k C^{-k} q^{k^2} \zeta^k \quad ,$$

with $C = a(\omega_2)/a(\omega_1)^\tau = -(-1)^{-\tau} = -e^{-\pi i\tau}$. Substituting,

$$T_0(u) = \sum_{k\in\mathbb{Z}} (-1)^k e^{\pi i\tau(k^2+k)} \cdot e^{2\pi iku} \quad .$$

Grouping the terms k and -k-1 for $k \geqslant 0$ gives

$$T_0(u) = \sum_{k\geqslant 0} (-1)^k q^{k(k+1)} \left\{ e^{2\pi iku} - e^{-2\pi iku - 2\pi iu} \right\},$$

hence finally

$$(2.11) \qquad T_0(u:L) = 2ie^{-\pi iu} \sum_{k\geqslant 0} (-1)^k q^{k(k+1)} \sin(\pi(2k+1)u) \quad .$$

The dependence of T_0 in the lattice L is through $q = e^{\pi i\tau} = \exp(\pi i\omega_2/\omega_1)$. This function has a zero at u = 0 and since we have taken d = 1 (and $\underline{d} \geqslant 0$) it must be the only theta function of its type (up to a multiplicative constant) by the theorem. It must in particular be proportional to σ. Coming back to $u = z/\omega_1$ we have

$$(2.11)' \qquad \sigma(z) = ce^{-\pi iz/\omega_1} \sum_{k\geqslant 0} (-1)^k q^{k(k+1)} \cdot \sin[\pi(2k+1)z/\omega_1] \quad .$$

We may introduce an order on the group of divisors of E as follows :

$$\underline{d} = \sum d_k(a_k) \geqslant 0 \quad \text{whenever} \quad d_k \geqslant 0 \text{ for all } k ,$$

and if more generally $\underline{d}$, $\underline{d}' \in \mathrm{Div}(E)$,

$$\underline{d} \geqslant \underline{d}' \quad \text{whenever} \quad \underline{d} - \underline{d}' \geqslant 0 \quad .$$

In particular if $\underline{d}$ is the divisor of a theta function θ, we see that $\mathrm{div}(\theta) \geqslant 0$ exactly when θ is <u>holomorphic</u>. Although the divisor of the zero function is not defined, we make the formal convention that $\mathrm{div}(0) \geqslant \underline{d}$ for all $\underline{d} \in \mathrm{Div}(E)$.

Then we define

$$L(\underline{d}) = \left\{ f \text{ L-elliptic: } \mathrm{div}(f) \geqslant -\underline{d} \right\}$$

for any divisor $\underline{d} \in \mathrm{Div}(E)$. Due to the preceding convention $0 \in L(\underline{d})$, and it is easily seen that $L(\underline{d})$ is a complex vector space (the notation L comes from the appellation <u>l</u>inear system, sometimes used for this space, or for the set of divisors of the form div(f) for $f \in L(\underline{d})$).

(2.12) <u>Corollary</u> (Riemann-Roch). <u>If</u> $\underline{d}$ <u>is a divisor on</u> E <u>of degree</u> $\geqslant 1$, <u>then</u> $\dim_{\mathbb{C}} L(\underline{d}) = \deg(\underline{d})$.

<u>Proof</u> : Let θ be any fixed theta function with $\mathrm{div}(\theta) = \underline{d}$ (we may suppose θ reduced, cf. (2.9)) and let (h,a) be the type of θ. Define then $\phi \longmapsto \phi/\theta$ on the space of theta functions of fixed type (h,a). Because ϕ and θ have same type, their quotient $f_\phi = \phi/\theta$ is an elliptic function, and

$$\mathrm{div}(f_\phi) = \mathrm{div}(\phi) - \mathrm{div}(\theta) = \mathrm{div}(\phi) - \underline{d} \quad .$$

We see that

$$\phi \text{ holomorphic} \Leftrightarrow \mathrm{div}(\phi) \geqslant 0 \Leftrightarrow \mathrm{div}(f_\phi) \geqslant -\underline{d} \quad ,$$

so that $\phi \longmapsto f_\phi = \phi/\theta$ establishes an isomorphism (of inverse $f \longmapsto f\theta$) between the d-dimensional space of holomorphic theta functions of type (h,a) (Th.(2.10)) and $L(\underline{d})$.

Let us now indicate how the theta functions are bound with the cohomology of the elliptic curve E, and in particular with the holomorphic line bundles on E.

To any L-theta function θ we associate a holomorphic line bundle $F_\theta \longrightarrow E$ on E as follows. Let (h,a) be the type of θ. We define a holomorphic action on the trivial line bundle on $\tilde{E} = \mathbb{C}$, compatible with the translations in the base by

$$(z,v)^{\omega} = (z+\omega, a(\omega)e^{\pi h(\omega)(z+\frac{1}{2}\omega)}v).$$

The quotient of $\tilde{E} \times \mathbb{C} \longrightarrow \tilde{E}$ (first projection) under this action of L gives the holomorphic line bundle $F_\theta \longrightarrow E$. By definition, we could as well write $F_{(h,a)}$ instead of F_θ. It is obvious that

$$z \longmapsto (z, \theta(z))$$

defines an equivariant section of the trivial line bundle on $\tilde{E}$, hence can be identified with a section of the quotient $F_\theta \longrightarrow E$. (We say section instead of <u>meromorphic</u> section.) If θ' is another section of $F_\theta \longrightarrow E$, θ' can be identified with an equivariant section of the trivial line bundle on $\tilde{E}$, hence with a theta function of type (h,a). Thus θ'/θ is elliptic and $\theta' = f\theta$ (with an elliptic function f). Thus $F_\theta = F_{\theta'}$. Moreover, if θ is a trivial theta function, $F_\theta \longrightarrow E$ has a holomorphic never vanishing section θ, so that it is holomorphically trivial (hence the denomination of trivial theta function). If θ and θ' are two theta functions with same divisor, their quotient must be a trivial theta function, so that F_θ and $F_{\theta'}$ are holomorphically equivalent line bundles on E. Again, we might as well note $F_{\underline{d}}$ instead of F_θ, and summing up, we see that $F_{\underline{d}}$ depends only on the class of the divisor $\underline{d}$ mod P(E). If we assume the basic existence (by no means trivial) theorem of existence of meromorphic sections for all (holomorphic) line bundles over a Riemann surface, it is easy to see that $\underline{d} \longmapsto F_{\underline{d}}$ defines a bijection between Div(E)/P(E) and the set of isomorphism classes of (holomorphic) line bundles over E. This set is naturally a group for the tensor product (corresponding to the usual product of cross sections θ) so that $\underline{d} \longmapsto F_{\underline{d}}$ is a group isomorphism (the line bundles are often called invertible bundles for the reason that among the holomorphic vector bundles with finite dimensional fibers, they are precisely those which are invertible for the tensor product). Explicitely, the inverse bundle of $F_{(h,a)}$ is $F_{(-h,1/a)}$ as follows from the formula giving the type of 1/θ in function of the type of θ. The theorem of Appel-Goursat (2.10) may be restated as follows : The holomorphic sections of the line bundle $F_{\underline{d}} \longrightarrow E$, when $\deg(\underline{d}) \geqslant 1$, make up a complex space of dimension equal to $\deg(\underline{d})$. (When $\deg(\underline{d}) = 0$, this dimension is not always 0 : indeed, the constants are sections of the trivial line bundle, and so we see that this dimension is one when $\underline{d}$ is principal. The correction to bring to the formula giving this dimension for small degrees is due to Roch. Riemann only gave the formula for divisors of degree sufficiently high, with an exact bound. With this in view, the Corollary (2.12) should be called Riemann's instead of Riemann-Roch's!)

We come finally to the cohomological interpretation of the theta functions and classes of divisors, by introducing the structure sheaf $\mathcal{O}$ of E (sheaf of germs of holomorphic functions on E), and its subsheaf of units $\mathcal{O}^{\times}$ (sheaf of germs of nowhere vanishing holomorphic functions). Let also $\underline{e}$ denote as usual the normalized exponential (of period 1) considered as sheaf homomorphism $\underline{e} : \mathcal{O} \longrightarrow \mathcal{O}^{\times}$. Its kernel is the constant sheaf $\underset{\sim}{\mathbb{Z}}$, and we have an exact sequence of sheaves

$$0 \longrightarrow \underset{\sim}{\mathbb{Z}} \longrightarrow \mathcal{O} \xrightarrow{\underline{e}} \mathcal{O}^{\times} \longrightarrow 1 \quad .$$

This exact sequence gives rise to the exact cohomology sequence

$$\begin{array}{ccccccccccc} H^0(E,\mathcal{O}) & \longrightarrow & H^0(E,\mathcal{O}^\times) & \longrightarrow & H^1(E,\underset{\sim}{\mathbb{Z}}) & \longrightarrow & H^1(E,\mathcal{O}) & \longrightarrow & H^1(E,\mathcal{O}^\times) & \longrightarrow & H^2(E,\underset{\sim}{\mathbb{Z}}) \longrightarrow 0 \\ \| & & \| & & & & & & & & \\ \mathbb{C} & \xrightarrow{\underline{e}} & \mathbb{C}^\times & & & & & & & & \end{array}$$

Because $\underline{e} : \mathbb{C} \longrightarrow \mathbb{C}^\times$ is surjective, we may cut the beginning of the above sequence and get the shorter exact sequence

$$(2.13) \qquad 0 \longrightarrow H^1(E,\underset{\sim}{\mathbb{Z}}) \longrightarrow H^1(E,\mathcal{O}) \longrightarrow H^1(E,\mathcal{O}^\times) \longrightarrow H^2(E,\underset{\sim}{\mathbb{Z}}) \longrightarrow 0.$$

We interpret each term. By Serre's duality $H^1(E,\mathcal{O})$ is $H^0(E,\Omega^1)$: the complex vector space $\Omega^1(E)$ of global holomorphic differentials on E. These differentials are all proportional to any non-zero of them (and all invariant under translations, as we shall see later) so that

$$H^1(E,\mathcal{O}) \cong \mathbb{C}\cdot\omega \cong \mathbb{C} .$$

Then $H^1(E,\underset{\sim}{\mathbb{Z}})$ is a lattice in $H^1(E,\mathcal{O})$ (the lattice of "integral" holomorphic differential forms) which can be identified with $L \hookrightarrow \mathbb{C}$ under the preceding isomorphism. The sequence (2.13) starts thus as follows : $0 \longrightarrow L \longrightarrow \mathbb{C} \longrightarrow \dots$. The last term $H^2(E,\underset{\sim}{\mathbb{Z}})$ is easily identified with $H^2(L,\mathbb{Z})$ (group cohomology of L with trivial action on $\mathbb{Z}$ because L has no torsion, acts freely on the contractible space $\mathbb{C}$ with quotient precisely E : Hurewitz Theorem). This group can also be interpreted as space of $\mathbb{Z}$-bilinear alternate integral forms on L. These forms are all proportional to the canonical one B :

$$H^2(L,\mathbb{Z}) \cong \mathbb{Z}\cdot B \cong \mathbb{Z} \quad .$$

The space $H^1(E,\mathcal{O}^\times)$ is the most interesting one. It can be computed by taking a covering of E with open discs U_α in $\mathbb{C}$ (mod L), each U_α being sufficiently small to avoid pairs $z, z+\omega$ $(\omega \in L - \{0\})$. Since all the U_α and their finite intersections $U_{\alpha_1\cdots\alpha_\ell} = \bigcap_{1\leq i\leq \ell} U_{\alpha_i}$ are contractible (by convexity), hence cohomologically trivial, the sheaf cohomology $H^1(E,\mathcal{O}^\times)$ is already the (sheaf-)Čech cohomology of the nerve of this covering (with variable coefficients in the groups $\mathcal{O}^\times(U_{\alpha_1\cdots\alpha_\ell})$). A 1-cochain for this cohomology is a function

$$s : (\alpha,\beta) \longmapsto s_{\alpha\beta} \in \mathcal{O}^{\times}(U_{\alpha\beta}) \text{ when } U_{\alpha\beta} \neq \emptyset$$

on the couple of indices (α,β) of the open intersecting discs U_α . A 1-cocycle is a 1-cochain s such that $\partial s = 0$, i.e.

$$\partial s : (\alpha,\beta,\gamma) \longmapsto s_{\beta\gamma}\, s_{\alpha\gamma}^{-1}\, s_{\alpha\beta} = 1 \text{ on } U_{\alpha\beta\gamma} = U_\alpha \cap U_\beta \cap U_\gamma \ .$$

In other words, $s_{\alpha\gamma} = s_{\alpha\beta} \cdot s_{\beta\gamma}$ on $U_{\alpha\beta\gamma}$ (when this triple intersection is non-empty). This is a compatibility condition on the transition functions $s_{\alpha\beta}$ defined on the non-empty $U_{\alpha\beta}$. A 1-cocycle is said to be a 1-coboundary, or more simply to be trivial, when there exists a map $\sigma : \alpha \longmapsto \sigma(\alpha) \in \mathcal{O}^{\times}(U_\alpha)$ with $s = \partial\sigma : (\alpha,\beta) \longmapsto \sigma(\beta)/\sigma(\alpha)$ on $U_{\alpha\beta}$. To each 1-cocycle s one can associate a holomorphic line bundle $F_s \longrightarrow E$ on E as follows. We glue together holomorphically the trivial line bundles on the U_α , $p_\alpha : U_\alpha \times \mathbb{C} \longrightarrow U_\alpha$ by means of the $s_{\alpha\beta}$:

$$\begin{array}{ccc} (z,v) & \longmapsto & (z, s_{\alpha\beta}(z)v) \\ U_{\beta\alpha} \times \mathbb{C} & \longrightarrow & U_{\alpha\beta} \times \mathbb{C} \\ p_\beta \downarrow & & p_\alpha \downarrow \\ U_\beta \cap U_\alpha & = & U_\alpha \cap U_\beta \end{array} .$$

The compatibility conditions on the $s_{\alpha\beta}$ (cocycle condition) ensure the coherence of these patchings. The line bundle F_s will precisely be holomorphically trivial when s is trivial (existence of a global holomorphic, nowhere vanishing cross-section), so we have an isomorphism

$$\begin{array}{ccc} H^1(E,\mathcal{O}^{\times}) & \xrightarrow{\ \simeq\ } & \text{group of holomorphic line bundles } /E \\ s & \longmapsto & (F_s \longrightarrow E) \end{array} .$$

We have already indicated the fact that this group is also isomorphic to the group of classes of divisors Div(E)/P(E) . The coboundary homomorphism

$$\partial : H^1(E,\mathcal{O}^{\times}) \longrightarrow H^2(E,\mathbb{Z}) = \mathbb{Z}\cdot B$$

is given on the line bundles precisely by

$$\partial : F_{(h,a)} \longmapsto (1/2i)\partial h = d \cdot B$$

(where d is the degree of any theta function of type (h,a), or equivalently, the degree of any meromorphic section of $F_{(h,a)} \longrightarrow E$). This justifies in some sense the terminology adopted for "coboundary ∂h" of the homomorphism $h : L \longrightarrow \mathbb{C}$. (More precisely, we should have defined the coboundary of h by

$$\partial h(\omega_1,\omega_2) = (1/2i)(h(\omega_1)\omega_2 - \omega_1 h(\omega_2)) \quad , \text{ but... ! })$$

If we identify $\mathbb{Z}\cdot B$ with $\mathbb{Z}$, the coboundary ∂ gives what is known under the name of Chern character $c : H^1(E,\mathcal{O}^{\times}) \longrightarrow \mathbb{Z}$: the Chern class c(F) of a holomorphic line bundle $F \longrightarrow E$ is computed (in theory at least) by taking the degree of the divisor of any meromorphic section of $F \longrightarrow E$. With these interpretations, the exact sequence (2.13) gives the (isomorphic) exact sequence

$$(2.13)' \quad 0 \longrightarrow L \longrightarrow \mathbb{C} \longrightarrow \underbrace{\text{group of line bundles } /E}_{\text{or } H^1(E,\mathcal{O}^{\times}), \text{ or } Div(E)/P(E)} \xrightarrow{deg} \mathbb{Z} \longrightarrow 0$$

The subgroup of $H^1(E,\mathcal{O}^{\times})$ consisting of line bundles of Chern class zero (or the subgroup of classes of divisors of degree zero) is thus isomorphic to $E = \mathbb{C}/L$. This is the previously defined (1.24) isomorphism. These line bundles are also called flat line bundles on E and are precisely those which are topologically trivial. So we restate the definition :

$$\begin{aligned} Pic(E) &= \text{"connected component" of } H^1(E,\mathcal{O}^{\times}) = \\ &= \text{group of holomorphic line bundles over E} \\ &\quad \text{which are topologically trivial (trivializable!)} . \end{aligned}$$

It is important and interesting to note that $H^1(E,\mathcal{O}^{\times})$ is also the group of invertible sheaves over E (Modules or $\mathcal{O}$-modules, locally free of rank one), by associating to the bundle $F \longrightarrow E$ the sheaf of germs of its holomorphic cross-sections. The point is that this sheaf can be defined without speaking of the line bundle (a real advantage in characteristic $\neq 0$!) as follows. If $s \in H^1(E,\mathcal{O}^*)$ is a coherent system of transition functions, the sheaf $\mathcal{F}_s$ is defined by $\mathcal{F}_s(U) = \Gamma(U,\mathcal{F}_s)$ = group (or ring) of holomorphic functions on U. The restrictions are twisted as follows:

$$\mathcal{F}_s(U_\alpha) \longrightarrow \mathcal{F}_s(U_{\alpha\beta})$$

$$f \longrightarrow s_{\beta\alpha}\cdot \text{restr}(f) \text{ to } U_{\alpha\beta} \quad .$$

Equivalent formulations are, when starting with a divisor $\underline{d} = \sum d_k(a_k)$, to put

$$\mathcal{F}_{\underline{d}}(U) = \Big\{ f:U \to \mathbb{C} \ : \ f(z) = (z-a_k)^{d_k} h(z) \text{ in neighbd. of } a_k \text{ if in } U, \text{ with } h \text{ holomorphic} \Big\},$$

and when starting with a theta function θ of type (h,a), to put

$$\mathcal{F}_\theta(U) = \Big\{ f:U+L \to \mathbb{C} \text{ holomorphic, satisfying } f(z+\omega) = a(\omega)e^{\pi h(\omega)(z+\frac{1}{2}\omega)} f(z) \text{ for } z \in U+L \Big\} .$$

The sheaf restrictions in these two definitions are the usual restrictions of functions.

(2.14) Complement 1. We show why it is sufficient to consider multiplicators of the form $\exp(A(\omega)z + B(\omega))$ when treating theta functions. Let indeed $F \longrightarrow E$ be any holomorphic line bundle over the elliptic curve $E = \mathbb{C}/L$. We use again the basic existence theorem of Riemann surface theory which gives the existence of a meromorphic section s of this bundle. We choose a lifting of this section to the trivial line bundle over the universal covering $\tilde{E} = \mathbb{C}$ of E. This section $\tilde{s}$ has the property that $(\tilde{s}'/\tilde{s})(z + \omega) - (\tilde{s}'/\tilde{s})(z)$ is holomorphic everywhere (entire function) in z for a fixed element ω of L. But $\tilde{s}'/\tilde{s}$ is a section of the canonical (cotangent) line bundle over E. As the tangent, as well as the cotangent line bundles over E, are trivial, this proves that $(\tilde{s}'/\tilde{s})(z + \omega) - (\tilde{s}'/\tilde{s})(z)$ is constant in z, hence $(\tilde{\mathfrak{s}}'/\tilde{\mathfrak{s}})'$ is an elliptic function. This is what we wanted to prove.

(2.15) Complement 2. It is possible to give more explicitely than has been done, the possible types of theta functions. Let θ be a theta function of type (h,a). We have seen that

a) $h : L \longrightarrow \mathbb{C}$ is an additive homomorphism,

b) $\partial h = 2id \cdot B$ where $d = \mathrm{degdiv}(\theta)$,

and a simple computation shows that $a : L \longrightarrow \mathbb{C}^\times$ satisfies

c) $a(\omega + \omega') = (-1)^{dB(\omega,\omega')} a(\omega)a(\omega')$ for $\omega, \omega' \in L$.

In particular, $|a| : L \longrightarrow \mathbb{R}_+^\times$ is a homomorphism, and this is what was needed to decompose $\theta = \theta_o \cdot \theta_{red}$ as after (2.7). As the proof of (2.8) shows, we can write $h(z) = (d/S)\bar{z} + \alpha \cdot z$ with a complex constant α. Moreover, if $n(\omega)$ denotes the biggest positive integer n such that $\omega/n \in L$ (for $\omega \in L$), we see that $(-1)^{dn(\omega)} a(\omega)$ is a homomorphism $\psi : L \longrightarrow \mathbb{C}^\times$ the absolute value of which must be of the form $\omega \longmapsto \mathrm{Re}(\beta\omega)$ for a suitable complex constant β. Summing up, the three above properties a), b), and c) imply that

$$h(z) = (d/S)\bar{z} + \alpha \cdot z \quad ,$$

$$a(\omega) = (-1)^{dn(\omega)} e^{\mathrm{Re}(\beta\omega)} \chi(\omega) \quad ,$$

with complex constants α, β, and a unitary character $\chi : L \longrightarrow \mathbb{U}$ ($\mathbb{U} = \mathbb{C}^1$ = unit circle in $\mathbb{C}$). Conversely, all such functions satisfy the three conditions a), b), c) and hence form the type of a theta function θ (with divisor of degree d).

(2.16) In Jacobi's theory, the variable $q = e^{\pi i \tau}$ where $\tau = \omega_2/\omega_1$ plays as important a role as τ itself. If we put $t = q^2$, the normalized exponential $\underline{e}$ gives an isomorphism

$$\underline{e} : \mathbb{C}/\mathbb{Z} \xrightarrow{\sim} \mathbb{C}^\times ,$$

whence also

$$\underline{e} : \mathbb{C}/\mathbb{Z} + \mathbb{Z}\tau \xrightarrow{\sim} \mathbb{C}^\times/t^{\mathbb{Z}} \quad ,$$

where $t^{\mathbb{Z}}$ denotes the multiplicative subgroup of $\mathbb{C}^\times$ generated by t :

$$t^{\mathbb{Z}} = \{t^n : n \in \mathbb{Z}\} \quad .$$

Obviously, $\mathrm{Im}(\tau) > 0$ is equivalent to $|t|<1$. Thus we could have defined an elliptic curve as being a quotient of the multiplicative group $\mathbb{C}^*$ by a discrete subgroup generated by an element t of modulus $|t| < 1$ (or $|t|>1$, but not $|t| = 1$). We shall see that this point of view is more suited to the study of p-adic elliptic curves.

3. Variation of the elliptic curve and modular forms

To compare different elliptic curves, we consider two lattices L and L' in $\mathbb{C}$. Then we put $E = \mathbb{C}/L$, $E' = \mathbb{C}/L'$ and compare them by means of holomorphic maps.

(3.1) Proposition. Let $\varphi : E \longrightarrow E'$ be any holomorphic map. If φ is not constant, it is induced by an affine linear transformation $z \longmapsto \alpha z + \beta$ and is surjective. In particular, if we define $C_\beta : E \longrightarrow E'$ by $\dot{z} \longmapsto \dot{\beta}$ (constant map) and $T_\beta : E' \longrightarrow E'$ by $\dot{z} \longmapsto \dot{z} + \dot{\beta}$ (translation by $\dot{\beta}$ in E'), then $\varphi = \psi + C_\beta = T_\beta \circ \psi$ and ψ is a homomorphism $E \longrightarrow E'$. (If ψ is not constant, it is surjective.)

Proof: We choose a lifting $\tilde{\varphi}$ to the universal coverings $\tilde{E} = \tilde{E'} = \mathbb{C}$, and note that $\tilde{\varphi}(z + \omega) - \tilde{\varphi}(z)$ must be an element of L' for every $\omega \in L$. By continuity in z of this expression, we see that this difference is independent of z and by derivation, $\tilde{\varphi}'$ must be an entire L-elliptic function. By (1.1.a) we see that $\tilde{\varphi}'$ must be constant, and consequently $\tilde{\varphi}(z) = \alpha z + \beta$ with two complex constants α and β. Obviously α must send L into L' : $\alpha \cdot L \subset L'$. When $\alpha \neq 0$, this is a serious restriction on L and L'. All assertions of the proposition follow from that.

(3.2) Corollary. If φ is a holomorphic map $E \longrightarrow E'$, is injective and such that $\varphi(0) = 0$, then it is a group isomorphism, and the lattices L, L' are homothetic.

Another proof of this corollary would be as follows. Let a', b' and $c' = a' + b' \in E'$. By (1.22) the divisor $(0)+(c')-(a')-(b')$ is principal, hence of the form div(f') for a function f' on E'. Let $f = f' \circ \varphi$. Its divisor is

$$\mathrm{div}(f) = (0) + (c) - (a) - (b) ,$$

where we have put $a = \varphi^{-1}(a')$, ...This principal divisor must also satisfy Abel's condition (apply (1.22.d) again), which gives $c = a + b$. This proves that φ^{-1} is a homomorphism, hence φ is an isomorphism (and is biholomorphic).

For τ in the upper half-plane $H = \{\tau \in \mathbb{C} : \mathrm{Im}(\tau) > 0\}$ we introduce the notation L_τ for the lattice $L_\tau = \mathbb{Z} + \mathbb{Z}\cdot\tau \subset \mathbb{C}$. Every lattice L has a direct basis ω_1, ω_2 hence is homothetic to (at least) one L_τ (take $\tau = \omega_2/\omega_1$). But if we choose a direct basis such that $|\omega_1|$ is minimum among the modules of the non-zero elements $\omega \in L - \{0\}$, and then ω_2 of module minimum among the elements of $L - \mathbb{Z}\cdot\omega_1$ (so as to have a direct basis), then we can say much more on $\omega_2/\omega_1 = \tau$. First, by construction, $|\tau| \geqslant 1$. Then also $|\omega_2 \pm \omega_1| \geqslant |\omega_2|$ by choice of ω_2. This leads to $|\tau \pm 1| \geqslant |\tau|$ and if $\tau = x + iy$ is decomposed in real and imaginary parts, $(x \pm 1)^2 + y^2 \geqslant x^2 + y^2$, hence $|x \pm 1| \geqslant |x|$. This means $|x| = |\mathrm{Re}(\tau)| \leqslant \frac{1}{2}$. Replacing ω_2 by $\omega_2 - \omega_1$ if necessary, we may assume

(3.3) $$\mathrm{Im}(\tau) > 0 \ , \ |\tau| \geqslant 1 \ , \ -\tfrac{1}{2} \leqslant \mathrm{Re}(\tau) < +\tfrac{1}{2} \ .$$

Again, replacing (ω_1, ω_2) by $(\omega_2, -\omega_1)$ if necessary, we may assume $\mathrm{Re}(\tau) \leqslant 0$ when $|\tau| = 1$ (cf. picture below).

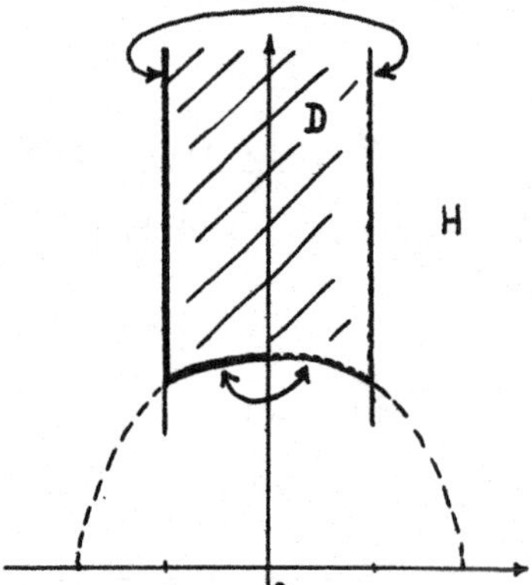

We denote by D the subset of the upper half-plane H defined by (3.3) and $\mathrm{Re}(\tau) \leqslant 0$ when $|\tau| = 1$. Then the lattices L_τ for $\tau \in D$ are a system of representants for the classes of homothetic lattices in $\mathbb{C}$. It can be checked directly that L_τ has only two direct bases satisfying the above conditions when $\tau \neq \underline{e}(1/3), i$, namely $(1,\tau)$ and $(-1,-\tau)$. In the case of the square lattice $\tau = i$, four direct bases of that kind are available : $(1,i)$, $(i,-1)$, $(-1,-i)$ and $(-i,1)$. Similarly in the case $\tau = \underline{e}(1/3) = \zeta$ (primitive $3^{\underline{d}}$ root of 1), there are six possible choices of such bases (hexagonal lattice). When we refer to the fundamental region D defined by (3.3), we always implicitely assume that $\mathrm{Re}(\tau) \leq 0$ when $|\tau| = 1$, as indicated just after (3.3).

Most functions encountered in Weierstrass' theory (first section) had homogeneity properties in the lattice. For example

$$\wp(\lambda z:\lambda L) = \lambda^{-2}\wp(z:L) \qquad (\lambda \in \mathbb{C}^{\times}) \quad ,$$

$$G_{2k}(\lambda L) = \lambda^{-2k}G_{2k}(L) \qquad \text{and}$$

$$\Delta(\lambda L) = \lambda^{-12}\Delta(L) \quad .$$

We introduce slightly different notations. When $\mathrm{Im}(\tau) > 0$, we put

$$\wp(z:\tau) = \wp(z:L_\tau) \quad ,$$

$$G_{2k}(\tau) = G_{2k}(L_\tau) \ , \quad \Delta(\tau) = \Delta(L_\tau) \ ,$$

always with $L_\tau = \mathbb{Z} + \mathbb{Z}\tau \subset \mathbb{C}$. Since Δ never vanishes, we can define the homogeneous function of degree zero

$$J(\tau) = g_2^3(\tau)/\Delta(\tau)$$

(remember $g_2 = 60G_4$ is homogeneous of degree -4 in the lattice, and we look at this function as function in $\mathrm{Im}(\tau) > 0$). It is important to note that by uniform convergence of

$$G_{2k}(\tau) = \sum{}'(m\tau + n)^{-2k} \qquad (k > 1)$$

for $\mathrm{Im}(\tau) \geq \varepsilon > 0$, we can evaluate the limit of this expression for $\tau = x + iy$ (x fixed) , and $y \to \infty$, by taking the limit of each term in the summation. All terms with $m \neq 0$ tend to zero, so that

$$\lim_{y\to\infty} G_{2k}(\tau) = \sum_{n\neq 0} n^{-2k} = 2\zeta(2k) \quad .$$

These constants can be computed explicitely (by comparing two expansions of cotg(z), and they can also be linked with the Bernoulli numbers: in particular, they can be seen to be $\neq 0$). The homogeneity properties of these functions as function of the lattice give very remarkable properties in the variable $\tau \in H = \{z \in \mathbb{C} : \mathrm{Im}(z) > 0\}$.

(3.4) Proposition. Let $\tau, \tau' \in H$ and $L_\tau, L_{\tau'}$ be the corresponding lattices. Then

a) $L_{\tau'} = L_\tau$ if and only if $\tau' = \tau + n$ with an integer n,

b) $L_{\tau'} \sim L_\tau$ if and only if $\tau' = (a\tau + b)/(c\tau + d) = \alpha(\tau)$ with a matrix $\alpha = \begin{pmatrix} a & b \\ c & d \end{pmatrix} \in SL_2(\mathbb{Z})$ (integral coefficients and determinant +1). In this case, the factor of proportionality is $c\tau + d$: $(c\tau + d)L_{\tau'} = L_\tau$.

Proof. The first part is easy. For the second, we take any $\lambda \in \mathbb{C}^\times$ such that $\lambda L_{\tau'} = L_\tau$. Since $\lambda\tau'$ and λ must belong to L_τ we can write

$$\begin{cases} \lambda\tau' = a\tau + b \\ \lambda = c\tau + d \end{cases} \qquad (*)$$

with $\alpha = \begin{pmatrix} a & b \\ c & d \end{pmatrix}$, an integral matrix. But conversely $L_{\tau'} = \lambda^{-1}L_\tau$ shows that there exists another integral matrix β with $\beta\cdot\alpha$ and $\alpha\cdot\beta$ of the form $\begin{pmatrix} 1 & n \\ 0 & 1 \end{pmatrix}$ (n an integer) by the first part. In particular, $\det(\alpha)$ must be an invertible integer. This proves $\det(\alpha) = \pm 1$.
A simple calculation shows that

$$\mathrm{Im}(\tau') = (ad - bc)\,|c\tau + d|^{-2}\mathrm{Im}(\tau) = \det(\alpha)\,|c\tau + d|^{-2}\mathrm{Im}(\tau) ,$$

hence $\det(\alpha) > 0$ by hypothesis $\tau, \tau' \in H$. Summing up, $\alpha \in SL_2(\mathbb{Z})$ as asserted. If conversely $\tau' = \alpha(\tau)$ with $\alpha \in SL_2(\mathbb{Z})$, the above computations show also that $(c\tau + d)L_{\tau'} = L_\tau$ q.e.d.

(3.5) Corollary. The holomorphic functions $G_{2k}(\tau)$ satisfy

$$G_{2k}(\alpha(\tau)) = (c\tau + d)^{2k}G_{2k}(\tau) , \; (k > 1)$$

for any $\alpha = \begin{pmatrix} a & b \\ c & d \end{pmatrix} \in SL_2(\mathbb{Z})$, and tend to a finite limit when τ tends to infinity on the imaginary axis.

This is just a reformulation of the homogeneity properties, in view of the proposition.

We introduce the following notation :

$$J(\alpha,z) = \text{Jacobian of } (\tau \mapsto \alpha(\tau)) \text{ at } \tau = z \quad , \quad (\alpha \in SL_2(\mathbb{Z})).$$

Thus, if $\alpha = \begin{pmatrix} a & b \\ c & d \end{pmatrix}$,

$$J(\alpha,\tau) = \left(\frac{a\tau + b}{c\tau + d}\right)' = (ac\tau + ad - ac\tau - cb)/(c\tau + d)^2$$

$$= (c\tau + d)^{-2} \quad .$$

(We hope that this function will never be confused with the modular invariant $J = g_2^3/\Delta$, especially since they do not depend on the same variables.) The corollary can be reformulated as

$$G_{2k}(\alpha(\tau)) = J(\alpha,\tau)^{-k} G_{2k}(\tau) \qquad (k > 1)$$

for $\tau \in H$ and $\alpha \in SL_2(\mathbb{Z})$. Also for example

$$J(\alpha(\tau)) = J(\tau) \quad \text{(by homogeneity of degree 0).}$$

When α is of the form $\begin{pmatrix} 1 & n \\ 0 & 1 \end{pmatrix}$ $(n \in \mathbb{Z})$, it gives an integral translation $\tau \mapsto \tau + n$ in H, having Jacobian 1, hence all the preceding functions are periodic of period one (even when the degree of homogeneity is $\neq 0$) and thus can be expanded in a Fourier series of the form

$$\sum_{-\infty}^{\infty} a_n e^{2\pi i n\tau} \quad .$$

Since the absolute value of $e^{2\pi i n\tau}$ for $\tau = iy$ purely imaginary, is $e^{-2\pi ny}$, we see that the G_{2k} can have no terms with index n negative (they are bounded when τ tends to ∞ on the imaginary axis). The same happens for $\Delta(\tau)$, and as consequence, $J(\tau)$ will have only finitely many non-zero Fourier coefficients of negative index n. The behavior of any holomorphic function with a Fourier expansion of the form $\sum_{n \gg -\infty} a_n e^{2\pi i n z}$ (finitely many $a_n \neq 0$ for $n < 0$) for $z = iy$ purely imaginary and $y \to \infty$ is given by the first term, with absolute value $|a_{n_0}| e^{2\pi |n_0| y}$ if $n_0 = -|n_0|$ is negative, and is the index of the first non-vanishing Fourier coefficient. These properties lead us to a general definition of modular forms.

(3.6) Definition. Let k be any integer. A (meromorphic) modular form of weight k is a meromorphic function f on the upper half-plane H, satisfying :

a) $f(\alpha(z)) = J(\alpha,z)^{-k} f(z)$ for $\alpha \in SL_2(\mathbb{Z})$ (whenever one member is defined!) ,

b) $|f(z)| \prec e^{\Lambda y}$ for some (positive) real constant Λ, when $y = \mathrm{Im}(z) \to \infty$.

The second property means that there exists a constant $C > 0$ such that $|f(z)| \leqslant Ce^{\Lambda y}$ for all y greater than some large y_0(depending on Λ, C and of course f). This condition ensures that the Fourier expansion of f (which exists by a)) has finitely many non-zero coefficients of negative index. The smallest index n corresponding to a non-zero Fourier coefficient, is called the order of f at infinity, and denoted by $\mathrm{ord}_\infty(f) = n_\infty$. The orders of f at all points $P \in D$ (fundamental region of H defined by (3.3)) or at all points $P \in H$ are defined as usual by considering the Laurent (or Taylor) expansion of f around P. For convenience, we put $\hat{D} = D \cup \{\infty\}$, so that we can speak of $\mathrm{ord}_P(f)$ when $P \in \hat{D}$, and $f \neq 0$.

The extent to which our examples G_{2k} exhaust the possibilities of holomorphic modular forms is tackled with a detailed study of the poles and zeros of general modular forms. We have indeed

(3.7) Proposition. Let $f \neq 0$ be a modular form of weight k. Then

$$\mathrm{ord}_\infty(f) + \tfrac{1}{2}\mathrm{ord}_i(f) + \tfrac{1}{3}\mathrm{ord}_\zeta(f) + \sum \mathrm{ord}_P(f) = k/6 ,$$

where $\zeta = \underline{e}(1/3)$ (primitive third root of 1) and the summation is extended over the points $P \in D - \{i,\zeta\}$ $(P \neq i,\zeta)$.

This proposition suggests strongly that we should adopt a new definition for the orders as follows :

$$v_i(f) = \tfrac{1}{2}\mathrm{ord}_i(f),\ v_\zeta(f) = \mathrm{ord}_\zeta(f)/3 \text{ and } v_P(f) = \mathrm{ord}_P(f)$$

for all other $P \in \hat{D}$. The formula could indeed be written then more simply

$$\sum_{P \in \hat{D}} v_P(f) = k/6 .$$

As for theta functions, we could show that the modular forms come from meromorphic cross-sections of suitable line bundles over $\Gamma\backslash H$ (space of orbits $\gamma(z)$, with $\gamma \in \Gamma = SL_2(\mathbb{Z})$) compactified by adding one point at infinity. These line bundles would be the $\otimes$-powers of the canonical one (cotangent bundle), so that a modular form of weight k comes from a k-differential on $\Gamma\backslash H \cup \{\infty\}$. Proposition (3.7) expresses the degree of their divisors.

Proof of (3.7): By hypothesis, $n_\infty = \mathrm{ord}_\infty(f) > -\infty$, so that f has no zero nor pole when $\mathrm{Im}(z) \geqslant R$ and R is large enough. Let D_R be the intersection of the fundamental domain D with $\mathrm{Im}(z) \leqslant R$. The idea is to integrate f'/f on the boundary ∂D_R of D_R. In case f has zeros or poles on the sides, f'/f will have simple poles on the sides, and the usual modification of the contour has to be made. By hypothesis f'/f has no pole on the top horizontal part of ∂D_R and if the modified contour is as shown on the picture (we suppose there that only one pole occurs on the vertical sides, and only one pole occurs on the portion $|z| = 1$, $-\frac{1}{2} < \mathrm{Re}(z) < 0$), it will enclose once each pole $\neq i, \zeta$ of f'/f in D. The two vertical sides cancel each other by periodicity of f. On the top horizontal portion HA (in this direction), we use the Fourier expansion of f (certainly valid for $\mathrm{Im}(z) \geqslant R$, because this region contains no pole of f):

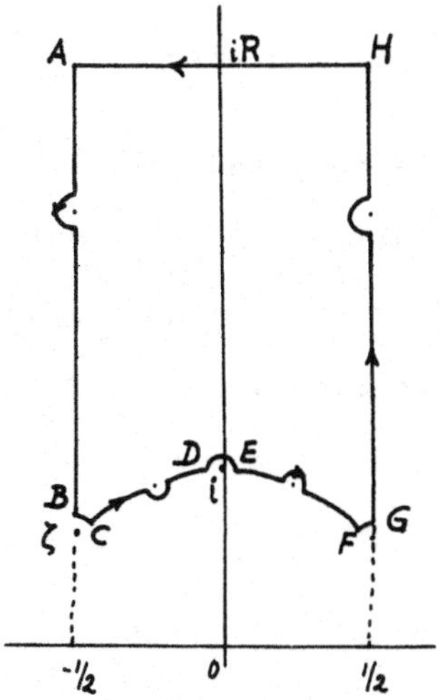

$$f(z) = a_{n_\infty} e^{2\pi i n_\infty z}(1 + \mathcal{O}(e^{2\pi i z})),$$

so that

$$f'(z)/f(z) = 2\pi i n_\infty + \mathcal{O}(e^{2\pi i z}),$$

or

$$\left|f'(z)/f(z) - 2\pi i n_\infty\right| = \mathcal{O}(e^{-2\pi \mathrm{Im}(z)})$$

for $\mathrm{Im}(z) \to \infty$. This shows that

$$\int_{HA} (f'/f)dz = -2\pi i n_\infty + \mathcal{O}(e^{-2\pi R}) \underset{R\to\infty}{\longrightarrow} -2\pi i n_\infty \ .$$

On the semi-circle (of radius $\varepsilon>0$ sufficiently small so as to enclose no pole of f'/f other than i), we get

$$\int_{DE} (f'/f)dz \longrightarrow -\tfrac{1}{2}(2\pi i \operatorname{ord}_i(f)) \quad \text{for } \varepsilon\to 0.$$

The arcs BC and FG can be treated similarly and give

$$\left(\int_{BC} + \int_{FG}\right)(f'/f)(z)dz \longrightarrow -\tfrac{2}{6}2\pi i\cdot \operatorname{ord}_\zeta(f) \text{ when } \varepsilon\to 0.$$

On CD, we suppose that the semi-circle chosen to avoid the pole of f'/f is the transform of that on EF (one of them having radius ε), under the mapping $z \longmapsto -1/z$ (this mapping transforms EF in DC). We compute the integral over EF by making the change of variable $\zeta = -1/z$, so that $z \longmapsto \zeta$ applies EF $\longrightarrow$ DC. Thus

$$(f'/f)(z)dz = (f'/f)(-1/\zeta)d(-1/\zeta) =$$
$$= (f'/f)(-1/\zeta)\zeta^{-2}d\zeta = (g'/g)(\zeta)d\zeta \ ,$$

with $g(\zeta) = f(-1/\zeta) = f(\begin{pmatrix}0 & -1\\ 1 & 0\end{pmatrix}\zeta) = \zeta^{2k}f(\zeta)$, because f is supposed to be a modular form of weight k. Hence

$$(g'/g)(\zeta)d\zeta = \left[2k/\zeta + (f'/f)(\zeta)\right]d\zeta \ .$$

Now we gather the two integrals over CD and EF of f'/f :

$$\left(\int_{CD} + \int_{EF}\right)(f'/f)(z)dz = \left(\int_{CD} + \int_{DC}\right)(f'/f)dz + \int_{DC} 2k\frac{d\zeta}{\zeta}$$

Only the integral of $2k\frac{d\zeta}{\zeta}$ over DC remains to be computed, or at least its limit when $\varepsilon\to 0$. For the limit, we can "rectify" the small semi-circle (of radius ε) to put CD on $|z| = 1$ (this does not change the value of the integral a bit !) and integrate from i to ζ . This is only the variation of the argument between i and ζ hence

$$\left(\int_{CD} + \int_{EF}\right)(f'/f)dz \longrightarrow 2k(2\pi i/12) = 2\pi i\cdot\frac{k}{6} \ .$$

The residue theorem gives on the other hand, the global result

$$\oint (f'/f)dz = 2\pi i \sum \operatorname{ord}_p(f) \quad ,$$

with a summation over the poles P of f'/f enclosed by the contour, i.e. all the poles of f'/f in D, distinct from i and ζ. The comparison between the explicit computation and the global result gives the announced formula q.e.d.

In the course of the proof, we have used the following result of function theory. Let g ($g = f'/f$ in our case) be an analytic function (in a neighborhood of the origin) having a simple pole at the origin, and let C_1 , C_2 be two continuously differentiable arcs starting at the origin, having respective tangents d_1 , d_2 with angle α. Then the integral of $g(z)dz$ on circles of radii ε between C_1 , C_2 and on circles of radii ε between d_1 , d_2 (both in the positive direction if α is positive) have same limit

$$(\alpha/2\pi)\,2\pi i\cdot \mathrm{Res}_{z=0}(g) \;=\; \alpha i\cdot \mathrm{Res}_{z=0}(g) \;.$$

We are mainly interested in holomorphic modular forms, so we define M_k to be the complex vector space of holomorphic modular forms f of weight k with $\mathrm{ord}_\infty(f) \geqslant 0$ (thus all $v_P(f)$ for $P \in \widehat{D}$ are $\geqslant 0$). We are going to show that all M_k are finite dimensional, and defining $d_k = \dim_{\mathbb{C}}(M_k)$, we shall show that we have the explicit formula

$$\sum_k d_k T^k = (1 - T^2)^{-1}(1 - T^3)^{-1} \in \mathbb{Z}[[T]] \;.$$

This formal series is called the Poincaré series of the graded module $M = \bigoplus_k M_k$.

(3.8) Consequences of (3.7). Let $f \in M_0$. Since the constants are in M_0 , $f - f(i)$ will be in M_0 . By construction, this function has a zero in i, so that the formula given by the proposition cannot hold. This proves $f - f(i) = 0$, hence $M_0 = \mathbb{C}$ is of dimension 1. Now if f, $\tilde{f} \in M_k$ have all same orders $\mathrm{ord}_P(f) = \mathrm{ord}_P(\tilde{f})$ $(P \in \widehat{D})$, their quotient $f/\tilde{f}$ will be in M_0 , hence constant. This shows that a function of M_k is determined up to a multiplicative constant by its orders at the $P \in \widehat{D}$. Obviously $M_k = \{0\}$ when the integer k is negative.

For $k = 1$, there is no possibility to satisfy (3.7) with integral orders, so that $M_1 = \{0\}$, $d_1 = 0$. For $k = 2$, there is only one possibility in (3.7): $\mathrm{ord}_\zeta(f) = 1$, all other orders being 0. This proves $d_2 = 1$, and all functions in M_2 vanish at ζ. We know that g_2 or G_4 is a holomorphic modular form of weight two, so that $M_2 = \mathbb{C}g_2 = \mathbb{C}G_4$ and g_2 vanishes at ζ with order one (simple zero)! Similarly for $k = 3$, we have only one possibility to satisfy (3.7): $\mathrm{ord}_i(f) = 1$ all other orders being zero. Hence $d_3 = 1$, $M_3 = \mathbb{C}g_3 = \mathbb{C}G_6$ and g_3 (or G_6) has only one simple zero located at $z = i$. Again for $k = 4$, there is only one possibility to satisfy (3.7): $\mathrm{ord}_\zeta(f) = 2$, all other orders being zero, so that $d_4 = 1$, $M_4 = \mathbb{C}g_2^2 = \mathbb{C}G_8$ and in particular, G_8, proportional to G_4^2, has a double zero at ζ (and no other zero). In full:

$$\textstyle\sum'(mz + n)^{-8} = c(\sum'(mz + n)^{-4})^2 \qquad (\mathrm{Im}(z) > 0),$$

a dream of youth, realized up to a constant (which can be determined explicitely). Finally there is only one possibility to satisfy (3.7) when $k = 5$: $\mathrm{ord}_i(f) = \mathrm{ord}_\zeta(f) = 1$ so that $d_5 = 1$ and $M_5 = \mathbb{C}g_2g_3 = \mathbb{C}G_{10}$. For larger k's, there are several possibilities. Since $\Delta \in M_6$ never vanishes in the upper half-plane, necessarily $\mathrm{ord}_\infty(\Delta) = 1$. Other functions in M_6 are g_2^3 and g_3^2. We can quote $f \in M_6$ with any prescribed simple zero $P \neq i, \zeta$ in D. It is sufficient to construct

$$g_3^2(P)\cdot g_2^3 - g_2^3(P)\cdot g_3^2 = ag_2^3 + bg_3^2 \quad .$$

The space M_6 is thus of dimension $d_6 = 2$, spanned by g_2^3 and g_3^2. But more generally, we have

(3.9) <u>Lemma</u>. <u>Let</u> $\varepsilon : M_k \longrightarrow \mathbb{C}$ <u>be the linear map defined by</u>

$$f \longmapsto \varepsilon(f) = \lim_{y\to\infty} f(iy) \quad ,$$

<u>and</u> $\Delta\cdot : M_k \longrightarrow M_{k+6}$ <u>be defined by multiplication by the function</u> Δ. <u>Then we have an exact sequence of complex vector spaces</u>

$$0 \longrightarrow M_k \xrightarrow{\Delta\cdot} M_{k+6} \xrightarrow{\varepsilon} \mathbb{C} \longrightarrow 0 ,$$

<u>and in particular</u> $d_{k+6} = d_k + 1$, <u>for</u> $k \geq 0$.

Proof: Because $\Delta \neq 0$, it follows that multiplication by Δ is an injective map. Put $M^o_{k+6} = \mathrm{Ker}(\varepsilon) \subset M_{k+6}$. Since $\mathrm{ord}_\infty(\Delta) = 1$, we see that the composite $\varepsilon \circ (\Delta \cdot)$ is zero. Conversely, if $f \in M_{k+6}$ is such that $\varepsilon(f) = 0$, i.e. f vanishes at infinity, it can be divided by Δ (this function will have the same orders at all $P \in D$ and order one less at ∞). Finally, $G_{2(k+6)}$ does not vanish at infinity (we could also take $g_2^\alpha g_3^\beta$ of weight $2\alpha + 3\beta$, with α and β chosen so that $2\alpha + 3\beta = k+6$, possible since 2 and 3 are relatively prime). The exactness gives $d_k - d_{k+6} + 1 = 0$.

From this lemma, and the explicit description of the M_k and d_k for $0 \leqslant k < 6$, we can check that

$$(1 - T^2)(1 - T^3) \sum_{k \geqslant 0} d_k T^k = 1 \quad .$$

We can also say the essentially equivalent result

$$d_k = \dim_{\mathbb{C}}(M_k) = \begin{cases} [k/6] \text{ (integral part of } k/6) \text{ if } k \equiv 1 \bmod 6 \\ [k/6] + 1 \text{ if } k \not\equiv 1 \bmod 6 \end{cases} .$$

Here, since these formulas are true for $0 \leqslant k < 6$, they will be true by induction for any k because $d_{k+6} = d_k + 1$.

We can state the general result.

(3.10) Theorem 1. Let $M = \bigoplus M_k$ be the graded algebra, sum of the spaces M_k of holomorphic modular forms of weight k (bounded when $z = iy$ and $y \to \infty$). Let on the other hand $\mathbb{C}[X,Y]$ be the algebra over $\mathbb{C}$, of polynomials in two indeterminates X,Y with the degree d defined by the two conditions $d(X) = 2$ and $d(Y) = 3$. Then

$$\Phi : \mathbb{C}[X,Y] \longrightarrow M \text{ defined by } X \longmapsto g_2 \text{ and } Y \longmapsto g_3$$

is an isomorphism of graded algebras. In other words, every holomorphic modular form, bounded when $z = iy$ is purely imaginary and $y \to \infty$, is a polynomial in g_2 and g_3 (or a polynomial in G_4 and G_6).

Here the degree d has been defined ad hoc, but if $\mathbb{C}[X',Y']$ is a polynomials algebra in two indeterminates, and if we look at the polynomials subalgebra generated by $X = X'^2$ and $Y = Y'^3$, this subalgebra will inherit the degree d from the natural degree of $\mathbb{C}[X',Y']$. In any case, $d(X^nY^m) = 2n + 3m$ by definition. Another way of saying the same thing would be to consider the graded algebras $\mathbb{C}[X]$, d' being defined as the double of the usual degree, and $\mathbb{C}[Y]$, d" being defined as the triple of the usual degree, and then $\mathbb{C}[X,Y] = \mathbb{C}[X]\otimes\mathbb{C}[Y]$ with the degree $d = d'\otimes d''$. The Poincaré series of $(\mathbb{C}[X],d')$ is obviously $1 + T^2 + T^4 + T^6 + \ldots = 1/(1 - T^2)$ because this algebra has one generator X^k in degree $d' = 2k$ and no non-zero element of odd degree. Similarly the Poincaré series of $(\mathbb{C}[Y],d'')$ is $(1 - T^3)^{-1}$. It is then well known (and easy to prove) that the Poicaré series of the tensor product of two graded algebras is the product of their Poincaré series, so that the Poincaré series of $(\mathbb{C}[X,Y],d)$ is $P(T) = (1 - T^2)^{-1}(1 - T^3)^{-1}$. This proves the theorem completely, but we give the classical proof as well.

<u>Proof</u> of Theorem 1. The elements of M_k for $0 \leqslant k < 6$ have been checked to be polynomials in g_2 , and g_3 . Since $\Delta = g_2^3 - 27g_3^2$ is also a polynomial in g_2 , and g_3 , induction applies by the lemma, and shows that any $f \in M_k$ (any integer k) is a polynomial in g_2 and g_3 , so Φ is surjective. We have to emphasize the fact that g_2 and g_3 satisfy no polynomial relation $P(g_2,g_3) = 0$ with $0 \neq P \in \mathbb{C}[X,Y]$. Decomposing P into homogeneous components (for the degree d) shows that we may assume P d-homogeneous to start with. If we had

$$P(g_2,g_3) = \sum_{2i+3j=a} c_{ij}g_2^ig_3^j = 0 ,$$

we could solve e.g. for g_2 and get

$$g_2^{i_0} = \sum_{j\neq 0} c'_{ij}g_2^ig_3^j \qquad \text{(after suitable division by } g_2^mg_3^n) ,$$

which is impossible, because all monomials in the right hand side contain a positive power of g_3 vanishing at $z = i$, and the left member vanishes only at $z = \zeta$ (with order i_0). This concludes the proof.

Now we call modular function a (meromorphic) modular form of weight 0. Our example is the modular invariant $J = g_2^3/\Delta$. Because $\mathrm{ord}_\infty(\Delta) = 1$, we have $\mathrm{ord}_\infty(J) = -1$, but J is holomorphic in H.

(3.11) Theorem 2. The modular invariant J defines a holomorphic bijection of the fundamental domain D defined in (3.3)(and after) onto the complex line $J : D \xrightarrow{\sim} \mathbb{C}$, which is conformal except at $z = i$ (ramification index 2) and at $z = \zeta$ (ramification index 3). This function is real on the sides of D and on the imaginary axis, with the normalizing values $J(\zeta) = 0$, $J(i) = 1$, $\lim\limits_{y\to\infty} J(iy) = \infty$. Finally, every modular function is a rational function of J.

Proof: J has a simple pole at $P = \infty \in \hat{D}$, so that the same will be true of any of the functions $J - \lambda$ (λ any complex value). In the formula of (3.7), only one compensation is possible, namely $v_P(J - \lambda) = 1$ for one (and only one) point $P \in D$. According to λ the following possibilities occur :

$$\mathrm{ord}_P(J - \lambda) = 1 \text{ with } P \neq i, \zeta \quad ,$$

$$\mathrm{ord}_i(J - \lambda) = 2 \text{ if } \lambda = J(i) = 1 \text{ (because } g_3(i) = 0) \quad ,$$

$$\mathrm{ord}_\zeta(J - \lambda) = 3 \text{ if } \lambda = J(\zeta) = 0 \text{ (because } g_2(\zeta) = 0) \quad .$$

This proves that J takes once and only once each complex value in the fundamental region D (not counting multiplicities). The second part of the theorem follows from :

$$G_{2k}(-\bar{z}) = G_{2k}(\bar{z}) = \sum{}'(m\bar{z} + n)^{-2k} = \overline{G_{2k}(z)} \quad ,$$

hence $J(-\bar{z}) = \overline{J(z)}$. In particular $J(z)$ will be real for $-\bar{z} = z$, i.e. for z purely imaginary. When $\mathrm{Re}(z) = \frac{1}{2}$, we have $-\bar{z} = z - 1$, so that the periodicity of J shows that it must also be real on the lines $\mathrm{Re}(z) = \pm\frac{1}{2}$. For $w = \begin{pmatrix} 0 & -1 \\ 1 & 0 \end{pmatrix}$ and for z on the unit circle, we have $-\bar{z} = w(z)$, so that $J(z) = J(w(z)) = J(-\bar{z}) = \overline{J(z)}$ must also be real.

Since J preserves the orientations, it must apply the part $Re(z) \leqslant 0$ of the fundamental region D onto the closed upper half-plane. From that (and the normalizing conditions), it can be reconstituted globally by means of the symmetry principle of Schwarz. Finally, let f be any (meromorphic) modular function. The conditions on f show that there must exist a meromorphic factorization F

$$\begin{array}{ccc} H\cup(i\infty) & \xrightarrow{\ f\ } & \mathbb{C}\cup(\infty) \\ J\downarrow & \nearrow_{F} & \\ \mathbb{C}\cup(\infty) & & \end{array}$$

with $\mathbb{C}\cup(\infty) = \mathbb{P}^1(\mathbb{C}) = S^2$ Riemann sphere. Since any meromorphic function on the Riemann sphere is a rational function, it follows that F is rational and that $f = F(J)$ is a rational function of J. q.e.d.

The second part of the theorem shows that J gives a conformal representation of $D \cap (Re(z) \leqslant 0)$ deleted of i, ζ onto $(Im(z) \geqslant 0) - \{0,1\}$, still with $J(\zeta) = 0$, $J(i) = 1$ (but no conformity at these points). Then using the symmetry principle of Schwarz on $Re(z) = -\frac{1}{2}$, 0 and then, -1, $+\frac{1}{2}$, ... will give the periodic function J in $D + \mathbb{Z}$. As the symmetry principle also applies along arcs of $\mathcal{D}$, it can be applied to $|z| = 1$, and inductively, it will eventually lead to a definition of J (by analytic extension) on H. Although this definition of J may be considered as more elementary than the one we have given (because it does not refer to elliptic curves in any way), it seems far from easy to derive from it the arithmetical properties of J which we are going to consider later. Also we note that the action of $SL_2(\mathbb{R})$ on H has a natural extension to $\mathbb{P}^1(\mathbb{R}) = \mathbb{R}\cup(i\infty)$ (considered as boundary of H) by fractional linear transformations. The transforms of $i\infty$ under $SL_2(\mathbb{Z})$ are the rational points $\mathbb{Q}\subset\mathbb{R}$ on the real axis, so that Δ must vanish at all these points (tend to 0 when we approach these points vertically from above) and the real axis is the natural boundary for the analytic extension of Δ (and of J).

(3.12) Corollary 1 (Little Picard Theorem). Let $f : \mathbb{C} \longrightarrow \mathbb{C}$ be an entire function. If $a \neq b \in \mathbb{C}$ are avoided by f : $f(\mathbb{C}) \cap \{a,b\} = \emptyset$, then f must be constant.

Proof: We may assume $a = 0$ and $b = 1$. Let $A \subset H$ be the set of ramification points of J : $A = J^{-1}\{0,1\}$. Then J defines a topological analytical covering $J : H - A \longrightarrow \mathbb{C} - \{0,1\}$ (local analytic isomorphism). By the monodromy principle ($\mathbb{C}$ is simply connected) $f : \mathbb{C} \longrightarrow \mathbb{C} - \{0,1\}$ can be lifted to the covering, $F : \mathbb{C} \longrightarrow H - A$. But Liouville's theorem (applied to $\exp(iF)$) shows that F must be constant (by continuity), and so must f be.

More important for us is the following

(3.13) Corollary 2. Every non-singular plane (projective) cubic curve $y^2t = x^3 + Axt^2 + Bt^3$, is isomorphic (as analytical group) to an elliptic curve $E = \mathbb{C}/L$, with a lattice $L \subset \mathbb{C}$.

Proof. The group law on this cubic is given by the condition that three points have sum zero if and only if they are on a line. If we make the transformation of variable $y \mapsto \tilde{y} = y/2$, we get an equation for $x, \tilde{y}$ in the Weierstrass form with $g_2 = -4A$ and $g_3 = -4B$. For this new equation, the non-singularity condition is $g_2^3 - 27g_3^2 \neq 0$. By the theorem, there exists a $z \in H$ such that $J(z) = g_2^3/(g_2^3 - 27g_3^2)$ and we define a first lattice $L = L_z$ in $\mathbb{C}$. Because J is homogeneous of degree 0 in the lattice, any homothetic lattice L would give the same value for J. We choose thus a complex constant $\lambda \neq 0$ such that moreover $g_2(\lambda L) = \lambda^{-4} g_2(L) = g_2 = -4A$. This is possible if $A \neq 0$: $\lambda^4 = -g_2(L)/(4A)$, and determines λ up to a power of $i = \sqrt{-1}$. In this case ($A \neq 0$ implies $J \neq 0$)

$$g_3^2(\lambda L) = g_2^3(\lambda L)(J(\lambda L) - 1)/(27J(\lambda L)) = g_3^2 ,$$

so that $g_3(\lambda L) = \pm g_3$. By homogeneity of g_3 of degree six in the lattice, changing λ to $i\lambda$ if necessary, will give the equality. For this lattice L', (1.11) gives an isomorphism $E = \mathbb{C}/L' \xrightarrow{\sim} X$ with the plane cubic curve $X : y^2t = 4x^3 - g_2xt^2 - g_3t^3$. In the case $A = g_2 = 0$, we note that the non-singularity condition implies $g_3 \neq 0$, so that one (at least) of the lattices λL_ζ $(\lambda \in \mathbb{C}^\times)$ will do. It is indeed sufficient to take for λ any sixth root of $g_3(L_\zeta)/g_3$.

(3.14) Corollary 3. Two elliptic curves $E = \mathbb{C}/L$ and $E' = \mathbb{C}/L'$ are isomorphic (as analytic spaces, or equivalently as analytic groups), if and only if $J(E) \equiv J(L) = J(L') \equiv J(E')$.

Proof: Indeed, two elliptic curves are isomorphic if and only if their lattices are homothetic. Now there are unique τ, τ' in the fundamental region D with $L \sim L_\tau$ and $L' \sim L_{\tau'}$. The result follows from the injectivity of J on D.

The corollary 2 is the basis of the algebraic study of elliptic curves. For example

(3.15) Application. Let X be the plane cubic of equation $(g_2^3 - 27g_3^2 \neq 0)$ $y^2t = 4x^3 - g_2xt^2 - g_3t^3$ and take any $\sigma \in \mathrm{Aut}(\mathbb{C})$ (not necessarily continuous). We look at σ as mapping $\mathbb{C}^2 \longrightarrow \mathbb{C}^2$ defined by $(x,y) \longmapsto (x^\sigma, y^\sigma)$. Then the image X_o^σ of $X_o = X \cap \mathbb{C}^2$ is the affine part of a plane cubic X^σ, and the two curves X and X^σ are isomorphic exactly when σ leaves $J = g_2^3/(g_2^3 - 27g_3^2)$ fixed.

It is obvious that X^σ has the equation $y^2t = 4x^3 - g_2^\sigma xt^2 - g_3^\sigma t^3$ (non-singular because X is assumed non-singular). Hence the invariant of X^σ is J^σ where J is the invariant of X. This proves the assertion. Note that if X is isomorphic to $\mathbb{C}/L$ and X^σ to $\mathbb{C}/L_\sigma$, we have in general no simple way of determining L_σ in function of L (we mean, no algebraic way of determining L_σ in function of L: the connection is given by the transcendental function J).

This application suggests that the field $\mathbb{Q}(g_2,g_3)\supset\mathbb{Q}(J)$ might be chosen to be equal to $\mathbb{Q}(J)$ for a special "model" of the equation of E (because $\mathbb{Q}(J)$ is the fixed field of the group of automorphisms $\sigma\in\mathrm{Aut}(\mathbb{C})$ satisfying $J^{\sigma} = J$). It is easy to show that this is true. If $g_3 \neq 0$, take $g = g_2' = g_3' = 27J/(J - 1)\in\mathbb{Q}(J)$. The curve

$$y^2t = 4x^3 - gxt^2 - gt^3$$

is isomorphic to E (it has the invariant $J = J(E)$). If $g_3 = 0$, any non-zero g_2 will do, for instance $g_2 = 1$, hence an equation

$$y^2t = 4x^3 - xt^2 = 4x(x - \tfrac{1}{2}t)(x + \tfrac{1}{2}t)$$

for E with coefficients in $\mathbb{Q} = \mathbb{Q}(J)$ ($J = 1$ in the case $g_3 = 0$).

(3.16) Remark 1. It is easy to check elementarily that G_6 vanishes at i and that G_4 vanishes at ζ. By definition

$$G_6(i) = \sum_{(m,n)\neq 0} (mi + n)^{-6} = \sum{}' i^6(-m + ni)^{-6}$$

$$= (-1)^3G_6(i).$$

Hence $2G_6(i) = 0$. Similarly since $\zeta^3 = 1$ and $\zeta^2 + \zeta + 1 = 0$ (minimal polynomial for ζ), we see

$$G_4(\zeta) = \sum_{(m,n)\neq 0} (m\zeta + n)^{-4} = \sum{}'\zeta^8(m\zeta^3 + n\zeta^2)^{-4}$$

$$= \sum{}' \zeta^2(m + n\zeta^2)^{-4} = \sum{}'\zeta^2(m + n(-1-\zeta))^{-4}$$

$$= \sum{}' \zeta^2(m-n - n\zeta)^{-4} = \zeta^2G_4(\zeta).$$

Hence $(1 - \zeta^2)G_4(\zeta) = 0$ and $G_4(\zeta) = 0$. This method does not give the orders of these zeros however, nor does it show that these "trivial" zeros are the only ones.

(3.17) Remark 2. Let $q = e^{\pi i\tau}$ be Jacobi's variable (cf. proof of (2.10)) and $t = q^2 = \underline{e}(\tau)$. Then $\tau \longmapsto t = \underline{e}(\tau)$ maps the region $-\frac{1}{2}\leqslant \mathrm{Re}(\tau) < \frac{1}{2}$, $\mathrm{Im}(\tau) > 0$ (one - one) onto the punctured unit circle $0<|t|<1$. Periodic meromorphic functions in H with period one correspond to meromorphic functions in the punctured unit circle under $\tau \longmapsto t$. The condition defining modular forms (3.6.b) means that the corresponding function on the punctured circle has at most a pole at the origin. The Laurent expansion of this function will thus be valid in a small punctured circle $0<|t|<r$ (r: distance from 0 to the first pole) and will be of the form

$$f(z) = F(t) = \sum_{n\geq n_0>-\infty} a_n t^n = \sum_{n\geq n_0} a_n e^{2\pi i n z} \quad .$$

A priori, the Fourier expansions of f are valid in regions $r_1<\operatorname{Im}(z)<r_2$ where no pole of f is in this band. But the growth condition (3.6.b) shows that r_1 can be chosen large enough to be able to take $r_2=\infty$. Another condition equivalent to (3.6.b) would be to require that $|f(z)|$ tend to a limit, finite or infinite when $\operatorname{Im}(z)\longrightarrow\infty$ (indeed F will at most have a pole at $t = 0$, as soon as we suppose that $|F(t)|$ tends to a limit, finite or infinite, for $t\longrightarrow 0$). The variable $t = q^2$ is better that $\tau\in H$ in the sense that if $t = \underline{e}(\tau)$ and $t' = \underline{e}(\tau')$ for some $\tau,\tau'\in H$, then the lattices L_τ and $L_{\tau'}$ are the same if and only if $t = t'$.

Automorphic forms have another interpretation in group theory. Let G be the Lie group $SL_2(\mathbb{R})$. Then the group of analytic automorphisms of the upper half-plane H is naturally identified with $G/(\pm 1)$, acting by fractional linear transformations. The stabilizer of $i\in H$ is the subgroup of matrices $g = \begin{pmatrix} a & b \\ c & d\end{pmatrix}$ such that

$$i = \frac{ai+b}{ci+d} = \frac{(ai+b)(d-ci)}{|ci+d|^2} = (i+ac+bd)|ci+d|^{-2} \quad ,$$

whence the conditions

$$c^2+d^2 = |ci+d|^2 = 1 \quad \text{and} \quad ac+bd = 0.$$

In G, they characterize the special orthogonal group $M = SO(2,\mathbb{R})$:

$$M = \left\{\begin{pmatrix} \cos\vartheta & \sin\vartheta \\ -\sin\vartheta & \cos\vartheta\end{pmatrix} ; \; 0\leq\vartheta<2\pi\right\} \text{ (stabilizer of i) } .$$

We introduce also the subgroups of G:

$$A = \left\{\begin{pmatrix} a & 0 \\ 0 & a^{-1}\end{pmatrix} : a\in\mathbb{R}_+^\times\right\}, \quad N = \left\{\begin{pmatrix} 1 & u \\ 0 & 1\end{pmatrix} : u\in\mathbb{R}\right\} .$$

We have then

$$A\cdot N = \left\{\begin{pmatrix} a & b \\ c & d \end{pmatrix} \in G : c = 0\right\} ,$$

and also obviously $G = M\cdot A\cdot N$ ($= N\cdot A\cdot M = G^{-1}$). This shows that the action of G on H is transitive :

$$G(i) = NAM(i) = NA(i) = N(i\mathbb{R}_+^\times) = \mathbb{R} + i\mathbb{R}_+^\times = H .$$

Thus we have an identification

(3.18) $$G/M \xrightarrow{\sim} H \text{ defined by } gM \mapsto gM(i) = g(i) = z .$$

In consequence we can also identify (with $\Gamma = SL_2(\mathbb{Z})$)

$$\Gamma\backslash G/M \xrightarrow{\sim} \Gamma\backslash H \text{ by } \Gamma gM \mapsto \Gamma(z) \text{ where } z = g(i).$$

In this way, $\Gamma\backslash G/M$ appears as set of the lattices of the form L_z . Although it is true that through the first identification every function f on H can be considered as function over G invariant under M, this is not useful for modular forms. We proceed differently. If f is a modular form of weight k (holomorphic on H, but no condition at $i\infty$), we define the corresponding function $F = F_f$ on the group:

(3.19) $$F(g) = J(g,i)^k f(z) \text{ where } z = g(i) \in H ,$$

and $J(g,i) = (ci + d)^{-2}$ for $g = \begin{pmatrix} a & b \\ c & d \end{pmatrix} \in G$.

With this definition, we see first that $J(m_\vartheta,i) = (\cos\vartheta - i\sin\vartheta)^{-2}$ $= e^{2i\vartheta}$ for $m_\vartheta = \begin{pmatrix} \cos\vartheta & \sin\vartheta \\ -\sin\vartheta & \cos\vartheta \end{pmatrix} \in M$, and then that

(3.20) $$\begin{aligned} F(gm_\vartheta) &= J(gm_\vartheta,i)^k f(gm_\vartheta(i)) = J(g,i)^k J(m_\vartheta,i)^k f(g(i)) = \\ &= F(g)e^{2ik\vartheta} = F(g)\chi_{2k}(m_\vartheta) , \end{aligned}$$

because $gm_\vartheta(i) = g(i) = z$, with the even character χ_{2k} of M (this character is the 2k-th power of the basic odd character χ of M defined by $\chi(m_\vartheta) = e^{i\vartheta}$). For a matrix $\gamma \in \Gamma$ we have

(3.21) $$\begin{aligned} F(\gamma g) &= J(\gamma g,i)^k \cdot f(\gamma(z)) = J(\gamma g,i)^k \; J(\gamma,z)^{-k} f(z) = \\ &= J(g,i)^k f(z) = F(g) . \end{aligned}$$

This shows that all the lifted functions F_f are functions over $\Gamma\backslash G$ (irrespective of the weight of f), the weight k appearing now in the character according to which F transforms along M. If we suppose that $f \in M_k^o$, i.e. f is a holomorphic modular form of weight k vanishing at $i\infty$, then the Fourier expansion of f will start with a term $e^{2\pi iz}$, hence will have the order of magnitude of $e^{-2\pi y}$ ($y = \mathrm{Im}(z)$) at most :

$$|f(z)| = \mathcal{O}(e^{-2\pi y}) \quad \text{for } y = \mathrm{Im}(z) \to \infty ,$$

and this will in turn give strong decreasing properties of $F = F_f$ along A, which ensure that F will be for example square summable on $\Gamma\backslash G$ (for the invariant measure coming from a Haar measure on G : because G is semi-simple, every Haar measure on G is bi-invariant). Thus the spaces M_k^o can be embedded in $L^2(\Gamma\backslash G)$, each of them being embedded in an isotypical component for the restriction to M of the right regular representation of G in $L^2(\Gamma\backslash G)$. The holomorphy condition on f gives a condition on F which can also be interpreted group theoretically, but this would lead us too far here. Let us treat an example. We take $f = G_{2k}$ (holomorphic of weight k), and we construct its corresponding function F on the group, which we still denote G_{2k} abusively . First, we note that

$$\sum{}'(mz+n)^{-2k} = \sum_{\ell\geq 1}\ \sum_{(c,d)=1} (c\ell z + d\ell)^{-2k} =$$

$$= \sum_{\ell\geq 1} \ell^{-2k} . \sum_{(c,d)=1} (cz+d)^{-2k} =$$

$$= \zeta(2k) \sum_{(c,d)=1} (cz+d)^{-2k} .$$

Thus, the problem is to lift $(cz+d)^{-2k}$ to the group. To simplify, we work with the normalized function $E_k(z) = G_{2k}(z)/(2\zeta(2k))$ (which satisfies $\lim_{y\to\infty} E_k(iy) = 1$).

(3.22) $$E_k(z) = \tfrac{1}{2} \sum_{(c,d)=1} (cz+d)^{-2k} .$$

Then we observe that all matrices $\gamma = \begin{pmatrix} a & b \\ c & d \end{pmatrix} \in \Gamma = SL_2(\mathbb{Z})$ have coefficients c,d prime to each other (the determinant condition gives Bezout's condition), and conversely, for each couple (c,d) of integers, prime to each other, Bezout's condition shows the existence of a couple of integers, say (a,b) such that the matrix $\begin{pmatrix} a & b \\ c & d \end{pmatrix}$ has determinant one. Now, it is easy to check that two matrices of Γ have same second lines if and only if they are congruent mod $\Gamma_\infty = \pm\begin{pmatrix} 1 & \mathbb{Z} \\ 0 & 1 \end{pmatrix}$. More precisely if γ and γ' have same second line c, d, then $\gamma' = \gamma_\infty \cdot \gamma$ with a matrix γ_∞ of the form $\begin{pmatrix} 1 & n \\ 0 & 1 \end{pmatrix}$ (n a rational integer) or of the form $\begin{pmatrix} -1 & n \\ 0 & -1 \end{pmatrix}$ (n as before). This gives in fact a bijection between the couples (c,d) of relatively prime integers, and the classes $\Gamma_\infty \gamma \in \Gamma_\infty \backslash \Gamma$. Using this, we can write

$$(3.23) \qquad E_k(z) = \tfrac{1}{2} \sum_{(c,d)=1} (cz + d)^{-2k} = \tfrac{1}{2} \sum_{\Gamma_\infty \backslash \Gamma} J(\gamma, z)^k .$$

On the group G, we have by definition

$$E_k(g) = J(g,i)^k E_k(g(i)) =$$

$$= \tfrac{1}{2} \sum_{\Gamma_\infty \backslash \Gamma} J(g,i)^k J(\gamma, g(i))^k = \tfrac{1}{2} \sum_{\Gamma_\infty \backslash \Gamma} J(\gamma g, i)^k .$$

If we put

$$L(g) = J(g,i) \quad ,$$

we find the very simple and suggestive formula

$$(3.24) \qquad E_k(g) = \tfrac{1}{2} \sum_{\Gamma_\infty \backslash \Gamma} L(\gamma g)^k \quad ,$$

showing that on the group, E_k is obtained in the cheapest way from L (except for the factor $\tfrac{1}{2}$) so as to be invariant on left under Γ. Because $L(ng) = L(g)$ for any $g \in G$ and $n \in (\pm 1)N$, it is indeed sufficient to take a sum of $L(\gamma g)$ over γ mod (left) $\Gamma \cap (\pm 1)N = \Gamma_\infty$ to obtain a (left-)Γ-invariant expression.

The fact that the hexagonal and square lattices are the only ones admitting more than two ("canonical") bases as constructed after (3.3) is equivalent to the fact that $\Gamma/(\pm 1)$ has only two finite subgroups of order 3 and 2 respectively, generated by $\begin{pmatrix} 1 & 1 \\ -1 & 0 \end{pmatrix}$ and

$\begin{pmatrix}0 & -1\\ 1 & 0\end{pmatrix}$ fixing ζ and i resp. From this and a careful study of the tesselation of H given by the transforms of the fundamental region D under $\Gamma/(\pm 1)$ one can conclude that this group is the free product of these two cyclic subgroups :

(3.23) $\qquad SL_2(\mathbb{Z})/(\pm 1) = C_2 * C_3 \qquad (C_n = \mathbb{Z}/n\mathbb{Z})$,

and also that Γ itself is amalgamated product of the subgroups of order 6 (generated by $\begin{pmatrix}1 & 1\\ -1 & 0\end{pmatrix}$) and of order 4 (generated by $\begin{pmatrix}0 & -1\\ 1 & 0\end{pmatrix}$) over their intersection of order two (generated by $\begin{pmatrix}-1 & 0\\ 0 & -1\end{pmatrix}$) :

(3.23)' $\qquad SL_2(\mathbb{Z}) = C_4 \underset{C_2}{*} C_6$.

(3.24) Proposition. For $y = \mathrm{Im}(z) \to \infty$ the following limit formulas are valid (uniformly in $x = \mathrm{Re}(z)$) :

$$\wp(az + b : L_z) \to -\pi^2/3 \quad \text{for a not integer} \ , \ (a,b \text{ real}),$$

$$\wp(b : L_z) \to \pi^2\left(\frac{1}{\sin^2 \pi b} - \frac{1}{3}\right) \text{ for b not integer} \ , \ (b \text{ real}),$$

$$(2\pi)^{-12}\Delta(z)\underline{e}(-z) \to 1 \ ,$$

$$J(z)\underline{e}(z) \to 1/1728 = (12)^{-3} = 2^{-6}.3^{-3} \quad .$$

Proof. By definition, we have

$$\wp(az + b : L_z) = (az + b)^{-2} + \sum_{m,n}{}'\left((az + b + mz + n)^{-2} - (mz + n)^{-2}\right).$$

The convergence of this series is uniform with respect to $\mathrm{Im}(z) > \varepsilon > 0$, so we compute the limit by taking the limit of each term. When a is not an integer, the terms $((a + m)z + (b + n))^{-2}$ tend to 0 for every m. Only remain the terms $(mz + n)^{-2}$ with $m = 0$ (and $n \neq 0$). The formula $\sum_{n>0} n^{-2} = \pi^2/6$ easily gives the first formula. When $a = 0$, but b is not an integer, only the terms $m = 0$ give a contribution,

$$\wp(b : L_z) \to b^{-2} + \sum{}'(b + n)^{-2} - \sum{}'n^{-2} =$$

$$= \sum (b - n)^{-2} - 2\zeta(2) = \pi^2 \sin^{-2}\pi b - \pi^2/3 \quad .$$

With the values $\zeta(4) = \pi^4/90$ and $\zeta(6) = \pi^6/945$, we find

$$g_2 = (2\pi)^4/12\cdot E_2 \quad , \quad g_3 = (2\pi)^6/216\cdot E_3 \quad ,$$

$$\Delta = (2\pi)^{12}/1728\cdot(E_2^3 - E_3^2) \quad , \quad J = E_2^3/(E_2^3 - E_3^2) \quad .$$

But the first coefficients of the t-expansion (Fourier expansion) of E_2 and E_3 may be computed easily by starting with the equalities

$$\pi \operatorname{cotg} \pi z = \pi i(t + 1)/(t - 1) = z^{-1} + \sum{}'[(z - n)^{-1} - n^{-1}],$$

and differentiating with respect to z three (resp. five) times both sides. We see then that

$$E_2(z) = 1 + 240t + \mathcal{O}(t^2) \; , \; E_3(z) = 1 - 504t + \mathcal{O}(t^2)$$

(a much stronger result, giving all Fourier coefficients explicitely will be derived in detail later, in (4.1)). This shows that

$$E_2^3 - E_3^2 = (3\cdot 240 + 2\cdot 504)\,t + \mathcal{O}(t^2) =$$

$$= 1728\,t + \mathcal{O}(t^2) \quad .$$

This finishes the proof of the proposition.

We turn to a closer sudy of the function λ of Legendre, and its dependence on the elliptic curve. By definition, if E is an elliptic curve, it has an equation of the form

$$E : y^2 = 4x^3 - g_2x - g_3 = 4\prod_{i=1}^{3}(x - \wp(\tfrac{1}{2}\omega_i))$$

and $\lambda = (e_3 - e_2)/(e_1 - e_2)$ with $e_i = \wp(\frac{1}{2}\omega_i)$ gives another possible equation for E of the form :

$$y^2 = x(x - 1)(x - \lambda) \quad \text{(cf. (1.13)) with } \lambda \neq 0 \, , 1 \, .$$

To bring a dependence in $z \in H$, we define explicitely

$$e_1(z) = \wp(\tfrac{1}{2} : L_z) \quad ,$$

$$e_2(z) = \wp(\tfrac{1}{2}z : L_z) \quad ,$$

$$e_3(z) = \wp(\tfrac{1}{2}(z+1) : L_z) \quad ,$$

and

$$\lambda(z) = (e_3(z) - e_2(z))/(e_1(z) - e_2(z)) \; .$$

If we take a direct basis ω_1 , ω_2 of L_z of the form

$$\omega_2 = az + b , \quad \omega_1 = cz + d ,$$

we shall obviously have $\frac{1}{2}\omega_1 \equiv \frac{1}{2}$ and $\frac{1}{2}\omega_2 \equiv \frac{1}{2}z \bmod L_z$, as soon as $a \equiv d \equiv 1$, $b \equiv c \equiv 0 \bmod 2$. We define the subgroup $\Gamma(2) \subset \Gamma = SL_2(\mathbb{Z})$ by the condition $\begin{pmatrix} a & b \\ c & d \end{pmatrix} \equiv \begin{pmatrix} 1 & 0 \\ 0 & 1 \end{pmatrix} \bmod 2$. In other words, $\Gamma(2)$ is the kernel of the natural homomorphism of reduction mod 2

$$SL_2(\mathbb{Z}) \longrightarrow SL_2(\mathbb{Z}/2\mathbb{Z}),$$

hence in particular is a normal subgroup of Γ with quotient isomorphic to the non-abelian group of order 6 (necessarily isomorphic to $\mathfrak{S}_3$). When α is in $\Gamma(2)$, we can thus write

$$\wp(\tfrac{1}{2}z: L_z) = \wp(\tfrac{1}{2}\omega_2: L_z) = J(\alpha, z)\, \wp(\tfrac{1}{2}\alpha(z):(cz+d)^{-1}L_z = L_{\alpha(z)}).$$

Hence

$$e_2(z) = J(\alpha, z) e_2(\alpha(z)) ,$$

and similarly for e_1 and e_3 . Coming back to λ , this gives the property

$$\lambda(\alpha(z)) = \lambda(z) \quad \text{for any } \alpha \in \Gamma(2) .$$

We say that λ is a <u>modular function</u> for the group $\Gamma(2)$. To have a full description of the behavior of this function under the fractional linear transformations coming from the full Γ , it is sufficient to look at a system of representatives of Γ mod $\Gamma(2)$. We have already observed that $w = \begin{pmatrix} 0 & -1 \\ 1 & 0 \end{pmatrix}$ and $n = \begin{pmatrix} 1 & 1 \\ 0 & 1 \end{pmatrix}$ are such that w is of order 2 and nw of order 3 in $\Gamma/(\pm 1)$; they will generate such a system of representatives. Consequently, it is sufficient to connect $\lambda(-1/z)$ and $\lambda(z+1)$ with $\lambda(z)$ to have the description of the behavior of λ under Γ. The first matrix corresponds to the change of basis

$$\omega_2 = -1 , \quad \omega_1 = z ,$$

or $\frac{1}{2}\omega_1 = z/2$, $\frac{1}{2}\omega_2 = -\frac{1}{2} \equiv \frac{1}{2}$, so that e_1 and e_2 are interchanged. This gives

$$\lambda(-1/z) = (e_3 - e_1)/(e_2 - e_1)(z) = 1 - \lambda(z) .$$

Under the second transformation,

$$\omega_2 = z + 1 , \quad \omega_1 = 1 , \quad \text{and} \quad \tfrac{1}{2}\omega_1 = \tfrac{1}{2} , \quad \tfrac{1}{2}\omega_2 = \tfrac{1}{2}(z + 1) ,$$

hence e_2 and e_3 are interchanged :

$$\lambda(z + 1) = (e_2 - e_3)/(e_1 - e_3)(z) = \lambda(z)/(\lambda(z) - 1) .$$

To sum up

(3.25) <u>Proposition</u>. <u>The function</u> λ <u>of Legendre is a modular function for the group</u> $\Gamma(2)$ (<u>normal subgroup of</u> Γ <u>of index</u> 6 <u>formed of all matrices congruent elementwise to the unit matrix mod</u> 2). <u>Under a fractional linear transformation of</u> Γ, λ <u>is transformed in one of the six functions</u> $\lambda , 1/\lambda , 1-\lambda , 1/(1-\lambda), 1 - 1/\lambda = (\lambda-1)/\lambda , \lambda/(\lambda-1)$, <u>and explicitely</u> $\lambda(-1/z) = 1 - \lambda(z)$, $\lambda(z+1) = \lambda/(\lambda-1)(z)$.

To give an explicit relation with J, it is necessary to study the limit of λ when $\mathrm{Im}(z) \to \infty$ in some detail. Since $\Gamma(2)$ contains the matrix $\begin{pmatrix} 1 & 2 \\ 0 & 1 \end{pmatrix}$ acting by a translation of 2 in H, the function λ is periodic of period 2. As before we put $q = e^{\pi i z}$, and the Laurent expansion of λ around $q = 0$ gives the Fourier expansion of λ (necessarily valid for $\mathrm{Im}(z) > 0$ because λ has no pole in the upper half-plane H). Using (3.24), we see that

$$(e_1 - e_2)(z) \longrightarrow \pi^2(1 - 1/3) + \pi^2/3 = \pi^2 \text{ for } \mathrm{Im}(z) \to \infty .$$

Also $(e_3 - e_2)(z) \longrightarrow 0$ for $\mathrm{Im}(z) \to \infty$, but we need more. Coming back to the definition of e_2 and e_3 by means of the $\wp$ function, and grouping together the corresponding terms in the expansion of $\wp$, (this cancels the convergence factors and gives an absolutely convergent series), we find

$$(e_3 - e_2)(z) = \sum_{m,n} \left[(\tfrac{1}{2} + \tfrac{1}{2}z + mz - n)^{-2} - (\tfrac{1}{2}z + mz - n)^{-2} \right] =$$

$$= \sum_{m,n} \left[((m+\tfrac{1}{2})z + \tfrac{1}{2} - n)^{-2} - ((m+\tfrac{1}{2})z - n)^{-2} \right] =$$

$$= \pi^2 \sum_{m} \left[\cos^{-2}(m+\tfrac{1}{2})\pi z - \sin^{-2}(m+\tfrac{1}{2})\pi z \right] ,$$

where we have used the classical expansion formula

$$\pi^2/\sin^2\pi z = \sum_{n\in\mathbb{Z}} (z - n)^{-2} \quad .$$

Using the duplication formulas for the trigonometric functions, we find that the difference $1/\cos^2\alpha - 1/\sin^2\alpha = -4\cos 2\alpha/\sin^2 2\alpha$.

We have thus obtained

$$(e_3 - e_2)(z) = -4\pi^2 \sum_m \cos(2m+1)\pi z/\sin^2(2m+1)\pi z =$$

$$= -8\pi^2 \sum_{m\geq 0} \cos(2m+1)\pi z/\sin^2(2m+1)\pi z \quad .$$

Furthermore $\cos\pi z = \frac{1}{2}(q + q^{-1})$ and $\sin\pi z = \frac{1}{2i}(q - q^{-1})$ hence

$$(e_3 - e_2)(z) = 16\pi^2 \sum_{m\geq 0} (q^{2m+1} + q^{-2m-1})/(q^{2m+1} - q^{-2m-1})^2 ,$$

$$\sum_{m\geq 0} \dots = \sum_{m\geq 0} q^{2m+1}(1 + q^{4m+2})/(1 - q^{4m+2})^2 =$$

$$= q(1 + q^2)/(1 - q^2)^2 + \sum_{m\geq 1} \dots = q + \mathcal{O}(q^2) \ .$$

This gives the desired formula

$$(e_3 - e_2)(z) = 16\pi^2 q + \mathcal{O}(q^2) ,$$

and so

(3.26) $$\lambda(z)q^{-1} = \lambda(z)e^{-\pi i z} \longrightarrow 16 = 2^4 \quad \text{for } \mathrm{Im}(z) \to \infty .$$

From this and the relation $\lambda(-1/z) = 1 - \lambda(z)$, we infer that $\lambda(iy) \to 1$ when $y\to 0$ (y real), and $\lambda(z + 1) = \lambda(z)/(\lambda(z) - 1)$ gives $\lambda(1 + iy) \to \infty$ when $y\to 0$ (ibid.).

We come to the explicit relation between λ and J. Any symmetric combination of the six functions appearing in (3.25) will be a modular function for Γ (the growth condition for $\mathrm{Im}(z) \to \infty$ is satisfied by what we have just seen). We construct the symmetric product of the translates by 1 of these functions :

$$f = (1 + \lambda)(1 + 1/\lambda)(1 + (1-\lambda))(1 + 1/(1-\lambda))(2 - 1/\lambda)(2\lambda-1)/(\lambda-1).$$

By (3.11), f is a rational function of J. Since f has no pole in H, it must be a polynomial in J, and more precisely since f has a simple pole at i

$$f = -\lambda^{-2}(1-\lambda)^{-2}(\lambda+1)^2(\lambda-2)^2(2\lambda-1)^2$$

is a linear function in J : $f = A + B\cdot J$. We determine these constants. The limits

$$q^{-2}f = e^{-2\pi iz}f(z) \longrightarrow -4\cdot 16 = -2^6 \quad ,$$

$$q^{-2}J = e^{-2\pi iz}J(z) \longrightarrow 1/1728 = 2^{-6}3^{-3} \; ,$$

show that $B = -3^3 = -27$. To determine A, we look at the special $z = i$. $g_3(i) = 0$ implies $e_1e_2e_3(i) = 0$, and quite generally $e_1 + e_2 + e_3 = 0$. If $e_1(i) = 0$, $e_2(i) = -e_3(i)$ gives $\lambda(i) = 2$. If $e_2(i) = 0$, similarly $\lambda(i) = -1$, and finally if $e_3(i) = 0$, $\lambda(i) = \frac{1}{2}$. In all cases $f(\lambda(i)) = 0$ and since $J(i) = 1$ this proves $A = 27$. We have obtained

$$f = 27 - 27J \quad \text{or} \quad J = 1 - f/27.$$

A small computation gives now

$$J = \frac{4}{27}\frac{(1-\lambda+\lambda^2)^3}{\lambda^2(1-\lambda)^2} \quad . \tag{3.27}$$

The relation between J and the e_i's is easily found now. From $e_1 + e_2 + e_3 = 0$, we get $e_i^2 = e_j^2 + e_k^2 + 2e_je_k$ for $\{i,j,k\} = \{1,2,3\}$, hence by summation $e_1^2 + e_2^2 + e_3^2 = -2(e_1e_2 + e_2e_3 + e_3e_1)$. Then (3.27) gives immediately

$$J = -4\,\frac{(e_1e_2 + e_2e_3 + e_3e_1)^3}{(e_1-e_2)^2(e_2-e_3)^2(e_3-e_1)^2} \quad . \tag{3.28}$$

Since $g_2 = -4(e_1e_2 + e_2e_3 + e_3e_1)$, this expression shows that

$$\Delta = 16(e_1-e_2)^2(e_2-e_3)^2(e_3-e_1)^2 = -16\prod_{i\neq j}(e_i-e_j) \; .$$

(it could, of course, have been derived more simply from it).

We draw two possible fundamental regions for $\Gamma(2)$ in H.

(3.29)

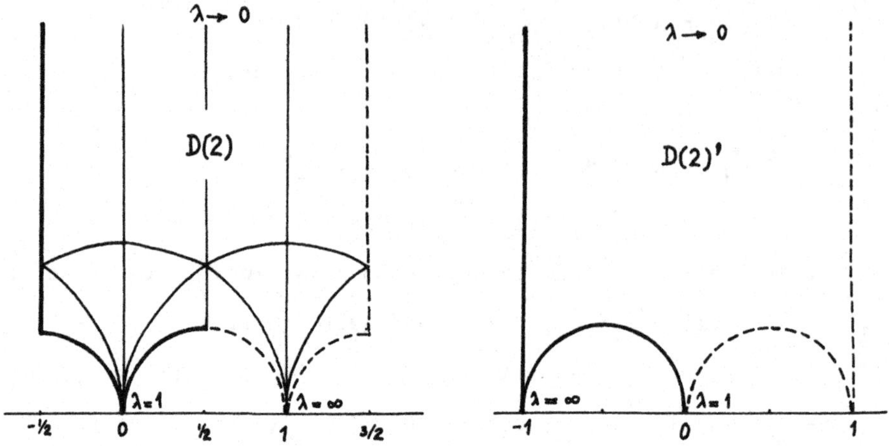

The behavior of a $\Gamma(2)$-invariant (say meromorphic) function in H, near $i\infty$ is made by introducing as usual the variable $q = e^{\pi i z}$. In the neighbourhood of 0 (in D(2) or D(2)'), we introduce the variable $q' = e^{\pi i z} = e^{-\pi i/z'}$ ($z' = -1/z$ near 0 means z near $i\infty$). When z" is in a neighbourhood of 1 (in D(2) or D(2)'), $z' = z''-1$ is near 0, so we take the variable $q'' = e^{\pi i z} = e^{-\pi i z'} = e^{-\pi i/(z''-1)}$.

(3.30) Definition. A modular function for $\Gamma(2)$ is a meromorphic function f in H satisfying

a) $f(\alpha(z)) = f(z)$ for all $\alpha \in \Gamma(2)$ (when one member is defined),

b) f has poles at most at the points $i\infty, 0, 1$ with reference to the Laurent expansion in the variables q, q', q'' respectively.

The function λ has these properties. Indeed, $\lambda(z) \to 0$ when $\mathrm{Im}(z) \to \infty$ (uniformly in Re(z)), $\lambda(z) \to 1$ when $z \to 0$ in D(2), so only the point 1 deserves verification. But there, we can use

$$\lambda(-1/(z-1)) = 1/(1 - \lambda(z)) \quad .$$

We could also look at (3.27) because J has simple poles (in the mentioned points) with respect to the variables $t = q^2$, $t' = q'^2$ and $t'' = q''^2$. The orders of a $\Gamma(2)$-modular function are thus well defined at all points of $\widehat{D(2)} = D(2) \cup \{i\infty, 0, 1\}$ (if the function does not vanish identically) , by means of the corresponding Laurent expansions. As in (3.7), integrating f'/f on the boundary $\partial D(2)$, making small semi-circles to avoid the zeros and poles of f and the points 0 and 1, we find

(3.31) $$\sum_{P \in \widehat{D(2)}} \mathrm{ord}_P(f) = 0 .$$

(This formula is simpler to establish than the corresponding one in (3.7) because we have no ramification points like we had there in ζ, i , and we take here only the case k=0.) Because λ is holomorphic and nowhere 0 in H, and because λ has a simple zero at $i\infty$, we see that λ and all the $\lambda - c$ ($c \in \mathbb{C}$) have a simple pole at 1. These functions must have one and only one zero in $\widehat{D(2)}$ to compensate in (3.31). This proves that λ gives a bijection

$$\lambda : D(2) \longrightarrow \mathbb{C} - \{0,1\} ,$$
$$\widehat{D(2)} \longrightarrow \mathbb{P}^1(\mathbb{C}) = \mathbb{C} \cup \{\infty\}.$$

Consequently (as in (3.11)), every modular function for $\Gamma(2)$ is a rational function in λ. We formulate an analogue of (3.11).

(3.32) <u>Proposition</u>. <u>The function</u> λ <u>of Legendre defines a conformal bijection of the fundamental region</u> D(2) <u>defined in</u> (3.29) <u>onto</u> $\mathbb{C} - \{0,1\}$. <u>This function is real on the imaginary axis and on the sides of</u> D(2)'. <u>The right half</u> $D(2)' \cap (\mathrm{Re}(z) > 0)$ <u>is mapped onto the upper half-plane with a continuous extension to the boundaries. Every</u> $\Gamma(2)$-<u>modular function is a rational function in</u> λ.

A couple of diagrams may help to clarify the situation.

(3.33)

$$\begin{array}{ccc} H & & \\ \downarrow \; \text{topological analytical covering} & & \\ \Gamma(2)\backslash H & \xrightarrow{\quad\lambda\quad} & \mathbb{P}^1(\mathbb{C}) - \{0,1,\infty\} \\ \downarrow \; \begin{matrix}\text{ramified covering} \\ \text{at } \zeta, i\end{matrix} & & \pi\downarrow \; \begin{matrix}\text{6-sheeted} \\ \text{ramified covering}\end{matrix} \\ \Gamma\backslash H & \xrightarrow[\quad J\quad]{} & \mathbb{P}^1(\mathbb{C}) - \{\infty\} \end{array}$$

The fibers of π have 6 elements except

$\pi^{-1}(0) = \{\lambda(\zeta),\ \lambda(w(\zeta)) = \lambda(\zeta+1)\}$ has only two elements, each corresponding to a ramification index 3,

$\pi^{-1}(1) = \{\lambda(i),\ \lambda((i-1)/i),\ \lambda(-1/(i-1))\}$ has only three elements, each corresponding to a ramification index 2.

In fact, π is the universal (maximal) covering of $\mathbf{P}^1(\mathbb{C})$ with respect to the three ramification conditions

index 3 above 0 ,

index 2 above 1 ,

index 2 above ∞ .

On the function fields of modular functions for Γ and $\Gamma(2)$ (both have been proved to be purely transcendental fields of degree one), π gives the injection $\mathbb{C}(J) \longrightarrow \mathbb{C}(\lambda)$. The group $\Gamma/\Gamma(2) \cong \mathfrak{S}_3$ acts naturally in $\mathbb{C}(\lambda)$ by field automorphisms over $\mathbb{C}$ (the group $\mathfrak{S}_3$ acts in the purely transcendental extension k(X) by k-automorphisms for <u>any</u> field k). The fixed field is precisely $\mathbb{C}(J)$ (by Lüroth's theorem the fixed field in k(X) is purely transcendental over a generator which can be taken $X^{-2}(1-X)^{-2}(1-X+X^2)^3$ of degree 6: compare with (3.27)). These facts show that $\mathbb{C}(\lambda)/\mathbb{C}(J)$ is a Galois extension of group isomorphic to $\mathfrak{S}_3$ and we can say that

$$\Gamma(2)\backslash H \longrightarrow \Gamma\backslash H$$

is a Galois covering of group $\Gamma/\Gamma(2) \cong \mathfrak{S}_3$.

Let us conclude this section by a few formulas, given without proof.

(3.34) $$\lambda(z) = 16q\prod_{n\geq 1}\left(\frac{1 + q^{2n}}{1 + q^{2n-1}}\right)^8$$

(3.35) $$\lambda(z) = 16\left(\frac{\sum_{n\geq 0} q^{(n+\frac{1}{2})^2}}{1+2\sum_{n\geq 1} q^{n^2}}\right)^4 = \left(\frac{\sum_{n\in\mathbb{Z}} q^{(n+\frac{1}{2})^2}}{\sum_{n\in\mathbb{Z}} q^{n^2}}\right)^4$$

Also the three singularities $\{0,1,\infty\}$ of λ make it possible to derive an inversion formula

$$z = i\frac{F(\frac{1}{2},\frac{1}{2}:1:1-\lambda)}{F(\frac{1}{2},\frac{1}{2}:1:\lambda)} \quad ,$$

where $F = {}_2F_1$ is the usual (Gauss') hypergeometric function.

4. Arithmetical properties of some (elliptic) modular forms

We are going to give explicitely the Fourier coefficients of the modular forms G_{2k} of weight k, or more precisely of the normalized $E_k = G_{2k}/(2\zeta(2k))$.

(4.1) Proposition. The modular form E_k of weight k ($k>1$) has the following Fourier expansion

$$E_k(z) = 1 + \gamma_k \sum_{N\geqslant 1} \sigma_{2k-1}(N)t^N \qquad (t = e^{2\pi iz} = \underline{e}(z))$$

where the constant γ_k is given by

$$\gamma_k = (-1)^k 4k/B_k = (-1)^k(2\pi)^{2k}/(\zeta(2k)(2k-1)!)$$

(B_k the Bernouilli number of index k), and $\sigma_\ell(N)$ is the arithmetical function defined by $\sigma_\ell(N)$ = sum of the ℓ-th powers d^ℓ of all the positive divisors $d>0$ of N.

Proof.. We start with the classical formula of function theory

$$\pi\cot g\pi z = z^{-1} + \sum_{n\geqslant 1}\{(z-n)^{-1} + (z+n)^{-1}\}$$

(note that the terms n and -n have to be grouped together to secure absolute and normal convergence on bounded sets of $\mathbb{C} - \mathbb{Z}$). By termwise derivation, we get

$$(\pi\cot g\pi z)' = -z^{-2} - \sum_{n\geqslant 1}\{(z-n)^{-2} + (z+n)^{-2}\}$$

$$= \sum_{n\in\mathbb{Z}} -(z+n)^{-2} \quad ,$$

and by induction

$$(\pi\cot g\pi z)^{(k)} = (-1)^k\cdot k!\sum_{n\in\mathbb{Z}}(z+n)^{-k-1} \qquad (k\geqslant 1) .$$

But the Fourier expansion of these functions is easily determined : with $q = e^{\pi iz}$, and $t = q^2 = \underline{e}(z)$, we get

$$\pi\cot g\pi z = \pi\cos\pi z/\sin\pi z = \pi i(q+q^{-1})/(q-q^{-1}) = \pi i(t+1)/(t-1),$$

whence

$$\pi\cot g\pi z = \pi i - 2\pi i/(1-t) = \pi i - 2\pi i(1+t+t^2+\dots) .$$

For the derivation, we note that $dt/dz = 2\pi it$, $d^k(t^n)/dz^k = (2\pi i)^k n^k t^n$. Hence

$$(\pi \operatorname{cotg} \pi z)^{(k)} = -(2\pi i)^{k+1} \sum_{n \geq 1} n^k t^n \quad , \text{ for } k \geq 1 \quad .$$

Comparison gives with mz instead of z (t^m instead of t)

$$\sum_{n \in \mathbb{Z}} (mz + n)^{-k-1} = (-1)^{k+1} (2\pi i)^{k+1}/k! \sum_{n \geq 1} n^k t^{mn} \qquad (m \neq 0) \ .$$

Also, replacing k by 2k - 1

$$\sum_{n \in \mathbb{Z}} (mz + n)^{-2k} = (2\pi i)^{2k}/(2k-1)! \sum_{n \geq 1} n^{2k-1} t^{mn} \quad .$$

Summing now over m (and treating the case m = 0 separately), we find

$$G_{2k}(z) = \sum{}' (mz + n)^{-2k} = 2\zeta(2k) + (2\pi i)^{2k} 2/(2k-1)! \sum_{m \geq 1} \sum_{n \geq 1} \ldots$$

because the sums with m and -m give the same result (cf. left hand-side). The double sum is $\sum_{m,n \geq 1} n^{2k-1} t^{mn}$ and its terms may be rearranged at will. We group together the terms with mn = N, a fixed integer $N \geq 1$, so that the coefficient of t^N is the sum $\sum_{n | N} n^{2k-1} = \sigma_{2k-1}(N)$ with the notations of the proposition. Summing up, we have obtained

$$E_k(z) = 1 + \gamma_k \sum_{N \geq 1} \sigma_{2k-1}(N) t^N \quad ,$$

with a coefficient

$$\gamma_k = (-1)^k (2\pi)^{2k} \zeta(2k)^{-1}/(2k-1)! = 2/\zeta(1-2k) \ .$$

Now we recall that the Bernouilli numbers may be defined by the expansion

$$\operatorname{cotg} z = z^{-1} - \sum_{k \geq 1} B_k 2^{2k} z^{2k-1}/(2k)! \quad .$$

The classical Weierstrass product formula

$$\sin z = z \prod_{n \geq 1} \left(1 - \frac{z^2}{n^2 \pi^2}\right) = z \prod_{n \geq 1} (n^2\pi^2 - z^2)/(n^2\pi^2) \quad ,$$

gives by logarithmic derivation

$$\operatorname{cotg} z = z^{-1} + 2 \sum_{n \geq 1} z/(z^2 - n^2\pi^2) =$$

$$= z^{-1} - 2z^{-1}\sum_{n\geq 1}\sum_{k\geq 1}(z/n\pi)^{2k} \quad .$$

Comparison gives then

$$2\zeta(2k)/\pi^{2k} = B_k 2^{2k}/(2k)!$$

and hence $\gamma_k = 4k(-1)^k/B_k$ as claimed.

We give the first few values of these numbers

$$B_1 = 1/6 \ , \quad B_2 = 1/30 \ , \quad B_3 = 1/42 \ , \quad B_4 = 1/30 \quad ,\ldots$$

$$\zeta(2) = \pi^2/6 \ , \quad \zeta(4) = \pi^4/90 \ , \quad \zeta(6) = \pi^6/945 = \pi^6/3^3\cdot 5\cdot 7 \quad ,\ldots$$

$$\gamma_1 = -24 \ , \quad \gamma_2 = 240 \ , \quad \gamma_3 = -504 \ , \quad \gamma_4 = 480 \quad ,\ldots$$

(4.2) Application. The relation $E_2^2 = E_4$ (both functions are normalized at $i\infty$, hence the proportionality constant is 1) gives the arithmetical relation of σ_3 and σ_7 . Indeed

$$E_2 = 1 + 240\sum_{n\geq 1}\sigma_3(n)t^n \quad ,$$

$$E_3 = 1 - 504\sum_{n\geq 1}\sigma_5(n)t^n \quad ,$$

$$E_4 = 1 + 480\sum_{n\geq 1}\sigma_7(n)t^n \quad ,$$

hence the explicit relation

$$\sigma_7(n) = \sigma_3(n) + 120\sum_{m=1}^{n-1}\sigma_3(m)\sigma_3(n-m) \quad .$$

Using (3.23) we may write $E_k = \tilde{\sum}(mz + n)^{-2k}$ with a summation extended over the couple of integers m,n satisfying

$$(m,n) = 1 \ , \ m > 0 \ , \ n > 0 \text{ if } m = 0 \ ,$$

hence the relation

$$\left(\tilde{\sum}(mz + n)^{-4}\right)^2 = \tilde{\sum}(mz + n)^{-8}$$

valid for $\mathrm{Im}(z) > 0$ (this relation was alluded to in (3.8)).
Similarly $E_3E_2 = E_5$ gives a relation between σ_3 , σ_5 , σ_9 .
By Theorem (3.10) every E_k is a polynomial in E_2 and E_3 hence σ_{2k-1} can be computed by means of σ_3 and σ_5 .

(4.3) Formulas. For reference we give explicitely some proportionality constants. With

$$E_2(z) = 1 + 240\sum_{n\geq 1}\sigma_3(n)t^n \qquad (t = \underline{e}(z)),$$

$$E_3(z) = 1 - 504\sum_{n\geq 1}\sigma_5(n)t^n \quad ,$$

we have

$$g_2 = 60\ G_4 = (2\pi)^4\cdot 2^{-2}\cdot 3^{-1}\cdot E_2 \quad ,$$

$$g_3 = 140\ G_6 = (2\pi)^6\cdot 2^{-3}\cdot 3^{-3}\cdot E_3 \quad ,$$

$$\Delta = (2\pi)^{12}\cdot 2^{-6}\cdot 3^{-3}(E_2^3 - E_3^2) \quad ,$$

$$J = E_2^3/(E_2^3 - E_3^2) = 2^2\cdot 3^{-3}\lambda^{-2}(1-\lambda)^{-2}(1-\lambda+\lambda^2)^3 \quad ,$$

and we introduce

$$j = 2^6\cdot 3^3\cdot J \qquad (2^6\cdot 3^3 = (12)^3 = 1728)$$

$$= 2^8\lambda^{-2}(1-\lambda)^{-2}(1-\lambda+\lambda^2)^3 \quad .$$

$$(\ j(\zeta) = 0\ ,\ j(i) = 1728\ ,\ j(i\infty) = \infty\ .)$$

(4.4) Theorem 1. The Fourier expansions of $(2\pi)^{-12}\Delta$ and of $j = 1728\cdot J$ have rational integral coefficients :

$$(2\pi)^{-12}\Delta(z) = \sum_{n\geq 1}\tau(n)t^n \quad , \qquad \tau(n)\in\mathbb{Z},\ \tau(1) = 1,$$

$$1728J(z) = 1/t + \sum_{n\geq 0}c(n)t^n \qquad (t = \underline{e}(z)),\ c(n)\in\mathbb{Z},$$

and in particular j is normalized by the condition of having residue $+1$ at $i\infty$.

Proof. Let us put $E_2 = 1 + 240\,U = 1 + 2^4\cdot 3\cdot 5\cdot U$ and $E_3 = 1 - 504\,V = 1 - 2^3\cdot 3^2\cdot 7\cdot V$, with two power series U, V in t and integral coefficients. By the above formulas, it is sufficient to show that

$$(1 + 2^4\cdot 3\cdot 5\cdot U)^3 \equiv (1 - 2^3\cdot 3^2\cdot 7\cdot V)^2 \bmod 2^6\cdot 3^3$$

in the sense that all coefficients of these series in t satisfy that congruence condition. Hence we have to check that

$$1 + 2^4\cdot 3^2\cdot 5\cdot U + 2^8\cdot 3^3\cdot 5^2\cdot U^2 + \ldots \equiv$$

$$\equiv 1 - 2^4 \cdot 3^2 \cdot 7 \cdot V + 2^6 \cdot 3^4 \cdot 7^2 \cdot V^2 \mod 2^6 \cdot 3^3 ,$$

or

$$2^4 3^2 5 \cdot U + 2^4 3^2 7 \cdot V \equiv 0 \mod 2^6 3^3$$

or still $5U + 7V \equiv 0 \mod 12$ $(= 2^2 3)$. But we observe that for any integer $n \geqslant 1$, the number $(n-1)n^2(n+1) = n^4 - n^2$ is divisible by 3 and 4 hence by 12, hence the congruence $n^4 \equiv n^2 \pmod{12}$ and a fortiori $n^5 \equiv n^3 \pmod{12}$. This gives $\sigma_3(n) \equiv \sigma_5(n) \pmod{12}$ and proves the desired congruence $5U + 7V \equiv 0 \pmod{12}$. Since $\tau(1) = 1$, the integrality property of j follows at once.

(4.5) Theorem 2. If we introduce the function of Dedekind

$$\eta(z) = e^{\pi i z/12} \prod_{n \geqslant 1} (1 - e^{2\pi i n z}) = q^{1/12} \prod_{n \geqslant 1} (1 - q^{2n})$$

then we have the following formulas

a) (Euler) $\eta(z) = e^{\pi i z/12} \sum_{n \in \mathbb{Z}} (-1)^n q^{n(3n+1)}$,

b) (Jacobi) $(2\pi)^{-12} \Delta(z) = \eta(z)^{24} = t \prod_{n \geqslant 1} (1 - t^n)^{24}$,

with the usual notations $q = e^{\pi i z}$ and $t = q^2 = \underline{e}(z)$.

Proof. The basis for this kind of theorem is (3.8) which shows that the dimension of the vector space M_6^o of modular forms of weight 6 vanishing at $i\infty$ is $\dim(M_6^o) = 1$, hence M_6^o is spanned by Δ: $M_6^o = \mathbb{C} \cdot \Delta$.

We put

$$g(z) = e^{\pi i z/12} \sum_{n \in \mathbb{Z}} (-1)^n e^{n(3n+1)\pi i z} =$$

$$= e^{\pi i z/12} \sum_{n \in \mathbb{Z}} \exp(\pi i 3 z n^2 + 2\pi i n(\tfrac{1}{2} z + \tfrac{1}{2})) ,$$

and moreover

$$\theta(z:\tau) = \sum_{n \in \mathbb{Z}} \exp(\pi i z n^2 + 2\pi i n \tau) , \quad \text{(Jacobi's } \theta_3).$$

Hence $g(z) = e^{\pi i z/12} \theta(3z : \tfrac{1}{2} + \tfrac{1}{2} z)$. We use now the classical summation formula of Poisson for which we recall the hypothesis.

If f is a continuous function on $\mathbb{R}$, $\sum_{n \in \mathbb{Z}} f(x + n)$ converges normally on every compact subset of $\mathbb{R}$ and $\sum_{n\in\mathbb{Z}} |\hat{f}(n)| < \infty$ ($\hat{f}$ denotes the Fourier transform of f), then we have $\sum f(x+n) = \sum \hat{f}(n)e^{2\pi inx}$.

For $f(x) = e^{\pi izx^2}$ (with $z = iy$, $y>0$ a real parameter), this gives

$$\sum \exp(\pi iz(x+n)^2) = \sqrt{i/z} \sum \exp(-(\pi i/z)n^2 + 2\pi inx) .$$

By analytic continuation for $z \in H$ and $x = \tau \in H$ this gives

$$\sum \exp(\pi iz(\tau+n)^2) = \sqrt{i/z}\ \theta(-1/z:\tau) .$$

Coming back to g :

$$g(-1/z) = e^{-\pi i/12z}\ \theta(-3/z : \tfrac{1}{2} - \tfrac{1}{2}z) =$$

$$= e^{-\pi i/12z}\sqrt{z/3i} \sum \exp(\pi iz/3(n + \tfrac{1}{2} - \tfrac{1}{2}z)^2) =$$

$$= \sqrt{z/3i} \sum_{m=2n+1} \exp[(\pi iz/12)(m^2 - 2m/z)] =$$

$$= \sqrt{z/3i} \sum_{m\text{ odd}} \exp(\pi im^2z/12 + \pi im/6) =$$

$$= \sqrt{z/3i} \sum_{m\text{ odd}} \exp(\pi im^2z/12)\cdot\cos(\pi m/6) .$$

When m is of the form 3m' (m' odd), $\cos(\pi m/6) = \cos(\pi m'/2) = 0$. Thus only the terms $m \equiv \pm 1 \pmod 6$ contribute in the summation, and since $m \mapsto -m$ interchanges these two classes, it is sufficient to take twice the sum over the class $m \equiv 1 \pmod 6$:

$$g(-1/z) = \sqrt{z/3i}\cdot 2\sum_{k\in\mathbb{Z}} \exp(\pi i(6k+1)^2z/12)\cdot\cos(\pi k + \frac{\pi}{6}) .$$

But $\cos(\pi k + \frac{\pi}{6}) = (-1)^k\sqrt{3}/2$, so that

$$g(-1/z) = \sqrt{z/i} \sum \exp(\pi i(36k^2 + 12k + 1)z/12) =$$

$$= \sqrt{z/i}\ e^{\pi iz/12} \sum \exp(\pi i3zn^2 + \pi izn) = \sqrt{z/i}\ g(z) .$$

We prove now that a similar functional equation holds for $\eta(z)$. Because H is simply connected and η never vanishes in H, we can choose a branch of $\log\eta$ in H :

$$(\log\eta)(z) = \pi iz/12 + \sum_{n\geq 1} \log(1 - e^{2\pi inz}) =$$

$$= \pi iz/12 + \sum_{n\geq 1}\sum_{m\geq 1} (-1/m)e^{2\pi inmz} .$$

Hence replacing z by -1/z :

$$\log \eta(-1/z) = -\pi i/12z - \sum_{n,m\geq 1} (1/m) e^{-2\pi i n m/z} \quad .$$

I claim that $\log \eta(-1/z) = \frac{1}{2}\log(z/i) + \log\eta(z)$ (this will prove $\eta(-1/z) = \sqrt{z/i}\,\eta(z)$), with the principal determination of log in $\mathbb{C} - i\mathbb{R}_+$ (hence the determination of $\sqrt{z/i}$ which takes the value +1 at z = i). We have to prove that

$$-\pi i/12z - \pi i z/12 + \sum_{n,m\geq 1} (1/m)(e^{2\pi i m n z} - e^{-2\pi i m n/z}) = \tfrac{1}{2}\log(z/i)$$

By summing first on n (the series is absolutely convergent), using the formula for the sum of a geometric series, we are led to a term

$$\frac{e^{2\pi i m z}}{1 - e^{2\pi i m z}} - \frac{e^{-2\pi i m/z}}{1 - e^{-2\pi i m/z}} = (i/2)(\mathrm{cotg}(\pi m z) + \mathrm{cotg}(\pi m/z)) \quad .$$

Remains to prove that the sum

$$-\pi i/12z - \pi i z/12 + (i/4) \sum_{n\neq 0} (\mathrm{cotg}(\pi n z) + \mathrm{cotg}(\pi n/z))/n$$

is equal to $\frac{1}{2}\log(z/i)$. To compute the sum, and prove this, we use the following trick of Siegel's. We consider the auxiliary function

$$\varphi(s) = \mathrm{cotg}(s)\cdot\mathrm{cotg}(s/z) \quad .$$

It has a double pole at s = 0 and simple poles for $s = n\pi$, $n\pi z$ when $0 \neq n \in \mathbb{Z}$. Since $\mathrm{cotg}(s) = s^{-1} - s/3 + \ldots$, and also $\mathrm{cotg}(s/z) = zs^{-1} - s/3z + \ldots$, we get

$$\mathrm{Res}_{s=0}\, \varphi(s)/(8s) = (-z/3 - 1/(3z))/8 = -z/24 - 1/24z \quad ,$$

$$\mathrm{Res}_{s=n\pi\neq 0}\, \varphi(s)/(8s) = (1/8n\pi)\mathrm{cotg}(n\pi/z)$$

$$\mathrm{Res}_{s=n\pi z\neq 0}\, \varphi(s)/(8s) = (1/8n\pi z)\mathrm{cotg}(n\pi z)\cdot z \quad .$$

This shows that the desiredsum is precisely $2\pi i \sum \mathrm{Res}\, \varphi(s)/8s$, with a summation extended over the whole s-plane. Let ν be a positive number, not an integer and P the parallelogram with vertices $\pm 1, \pm z$. Then

$$\oint_{\partial(\nu\pi P)} \varphi(s)/(8s)ds = \oint_{\partial P} \varphi(\nu\pi t)/(8t)dt$$

tends to the desired sum when $\nu \to \infty$. Since $t \mapsto \varphi(\nu\pi t)$ is uniformly bounded on ∂P for $\nu \mapsto \infty$ and since this function has the limits 1 on two opposite sides, -1 on the other opposite sides (vertices not included), the limit is easily estimated :

$$\lim_{\nu\to\infty} \oint_{\partial P} \varphi(\nu s)/(8s)ds = \tfrac{1}{4}\left(\int_1^z + \int_{-1}^z\right)(ds/s) =$$

$$= \tfrac{1}{4}(\log z - \log 1 + \log z - \log -1) = \tfrac{1}{2}\log(z/i)$$

with the principal determination of log in $\mathbb{C} - i\mathbb{R}_+$.
The conclusion of the proof of the theorem is now easy. Indeed, g^{24} as well as η^{24} have Fourier series starting with $t = q^2$ and are modular forms of weight 6 (indeed $SL_2(\mathbb{Z})$ is generated by $\begin{pmatrix}0 & -1\\ 1 & 0\end{pmatrix}$ and $\begin{pmatrix}1 & 1\\ 0 & 1\end{pmatrix}$). This characterizes Δ completely because $M_6^0 = \mathbb{C}\cdot\Delta$. More precisely, $(g - \eta)/g$ is a modular function vanishing at $i\infty$. Since it is holomorphic everywhere on H, it must be identically zero, this proves $g = \eta$, and finishes the proof of the theorem.

This theorem gives a relatively easy way of computing the first coefficients $\tau(n)$:

$\tau(1) = 1$, $\tau(2) = -24$, $\tau(3) = 252$, $\tau(4) = -1472$, ...

Ramanujan conjectured and Mordell proved (1920) that moreover

(4.6) $$\sum_{n\geq 1}\tau(n)n^{-s} = \prod_{p \text{ prime}} (1 - \tau(p)p^{-s} + p^{11-2s})^{-1}$$

for $Re(s) > 13/2$. The proof of this has now become a standard application of the theory of Hecke operators in modular forms. An equivalent way of saying (4.6) is

(4.6)' $\tau(mn) = \tau(m)\tau(n)$ when $(m,n) = 1$, $m \geq 1$, $n \geq 1$,

$\tau(p^{k+1}) = \tau(p)\tau(p^k) - p^{11}\tau(p^{k-1})$, p prime, $k \geq 1$.

These Ramanujan coefficients satisfy many famous congruences, e.g. except for a finite number of n, they are all divisible by 691 (although Ramanujan knew that for $n \leq 5000$ the $\tau(n)$ are never divisible by 691, he was lead to conjecture the preceeding result, which was later proved by Watson). Actually, Walfisz proved that the $\tau(n)$ are divisible by $2^5 \cdot 3^2 \cdot 5^2 \cdot 7 \cdot 691$ for nearly all n. Some conjectures remain open. For example, Ramanujan conjectured that $|\tau(p)| < 2p^{11/2}$. Deligne could recently show (1969) that this would indeed be the case if the general conjectures of A. Weil on the form of the zeta function of algebraic varieties over finite fields were proved to be true. Lehmer conjectures that $\tau(n) \neq 0$ for all n : this has been proved at least for $n < 10^{15}$. (For all these questions, we refer to J.-P. Serre : Une interprétation des congruences relatives à la fonction τ de Ramanujan, Séminaire Delange-Pisot-Poitou, 1967/68, n°14.)

The Fourier coefficients c(n) of j have also some interesting (and intriguing) arithmetical properties. Their growth is wild :

$$c(0) = 744, \; c(1) = 196884, \; c(2) = 21493760, \; \ldots$$

For example :

if n is divisible by 2^a, then c(n) is multiple of 2^{3a+8},
" " 3^b, " " 3^{2b+3},
" " 5^c, " " 5^{c+1}, ...

CHAPTER TWO

ELLIPTIC CURVES IN CHARACTERISTIC ZERO

Many properties of (complex) elliptic curves are purely algebraic in the sense that they could also be derived in the field generated over the rationals by the two numbers g_2 and g_3 , or perhaps in its algebraic closure. It is thus sufficient to work in a fixed field of characteristic zero (algebraically closed if needed). Our first aim will be to show that any non-singular (projective) plane cubic can be given in Weierstrass' normal form. The study of the differentials over the curve (and their classification) will also be made purely in algebraic terms.

It is true that by Lefschetz' principle the case of the complex field is crucial, and we can always reduce algebraic geometrical problems in characteristic zero to the same problems over $\mathbb{C}$. However, it is not always useful to do that. In particular, when the coefficients g_2 , g_3 are rational, it may be very interesting to look at the points of the curve $y^2 = 4x^3 - g_2x - g_3$ with coordinates in the p-adic fields $\mathbb{Q}_p$ (or algebraic extensions of these). It is very striking how Jacobi's theory of theta functions can be made here if the absolute invariant j is not a p-integer (i.e. $j \notin \mathbb{Z}_p$). Thus we conclude the chapter with a brief description of Tate's elliptic curves.

We shall assume that the reader is acquainted with the following facts of algebra :

a) If A is a noetherian ring, so is A[X] (Hilbert's basis theorem), and if A is a unique factorization domain, so is also A[X] (X is an indeterminate).

b) Notion of integral element over a ring and simple properties (sum and product of integers are integers, transitivity).

c) Notion of valuation (with value group contained in $\mathbb{R}$), and existence of extensions of a given valuation, corresponding to a field extension.*)

A certain familiarity with the p-adic fields $\mathbb{Q}_p$ (p a prime) and metric spaces with an ultra-metric distance, might also help in the last sections, although this is not absolutely required.

In the section on analytic p-adic functions, some propositions are proved for their interest and are not used after. Such are (4.2), (4.3) and (5.11).

---

*) Valuations are considered additively $v: K^{\times} \longrightarrow \mathbb{R} = \mathbb{R}^{+}$ whereas absolute values are considered multiplicatively $|...|_v : K^{\times} \longrightarrow \mathbb{R}_{+}^{\times}$.

1. Algebraic varieties and curves

Let k be any field and (P_i) a family of polynomials of $k[X_1,\dots,X_n]$ (where n is a strictly positive fixed integer). For any extension field K of k, we define the set $V_K \subset K^n$ as set of common zeros $x = (x_1,\dots,x_n) \in K^n$ of the family (P_i) : $P_i(x) = 0$ for all indices i. If $x \in V_K$, it is obvious that $P(x) = 0$ for every polynomial P of the ideal I generated by the P_i in $k[X_1,\dots,X_n]$. By Hilbert's basis theorem (the ring $k[X_1,\dots,X_n]$ is noetherian), the ideal I is generated by a finite number of polynomials, so we may suppose that the family (P_i) is finite to start with. We may then replace k by the field generated over the prime field by the finite number of coefficients occuring in the finite number of polynomials P_i , and thus suppose that k is finitely generated over the prime field. If k is of characteristic 0, we are thus reduced to the case of a field k finitely generated over $\mathbb{Q}$, and all such fields can be embedded in the field $\mathbb{C}$ of complex numbers ($\mathbb{C}$ is algebraically closed, and its transcendance degree over $\mathbb{Q}$ is not even denumerable). Thus if the characteristic is 0, we could also have started with $k = \mathbb{C}$: this is <u>Lefschetz' principle</u>. We come back to the general case, and note that the set V_k (or V_K) may very well be empty (e.g. when the ideal I is the whole ring $k[X_1,\dots,X_n]$ in which case we say that the ideal is not strict). But when K is algebraically closed we have

(1.1) <u>Theorem 1</u> <u>(Hilbert's Nullstellensatz)</u>. <u>Let</u> I <u>be a strict ideal of the ring</u> $k[X_1,\dots,X_n]$. <u>Then there is a common zero</u> $x = (x_i)$ <u>in a finite algebraic extension</u> K <u>of</u> k. <u>In other words</u> V_K <u>is not empty if</u> K <u>is an algebraic closure of</u> k.

Proof. The proof is made by using essentially the notion of integral element over a ring. First part : we show that if I is a maximal (strict) ideal of $k[X_1,\dots,X_n]$, then the field $K = k[X_1,\dots,X_n]/I = k[x_1,\dots,x_n]$ is algebraic over k. This is done by induction over n, the case n = 1 being trivial, because $k[x]$ is not a field if x is transcendental over k. We note then that

$$k(x_1)[x_2,\dots,x_n] = k[x_1,\dots,x_n] ,$$

because the right member is a field. By induction hypothesis, the n - 1 elements $x_2,\dots,x_n$ are algebraic over the field $k(x_1) = k'$, hence there is a polynomial $P(x_1) \in k[x_1]$ so that all $P(x_1)x_i$ for $i = 2,\dots,n$, are integral over $k[x_1]$ (this P is a common denominator for the x_i's). If now $f \in k(x_1) \subset k[x_1,\dots,x_n]$, a suitable positive integer $N = N(f)$ will be such that

$$P^N(x_1)\cdot f \in k[x_1]\,[P(x_1)x_2,\dots,P(x_1)x_n] ,$$

hence integral over $k[x_1]$ (by transitivity of the notion of integrality) This implies $P^N f \in k[x_1]$, or $f \in k[x_1, P(x_1)^{-1}]$ and finally

$$k(x_1) \subset k[x_1, P(x_1)^{-1}],$$

and so x_1 is algebraic over k . Hence the result by transitivity of the algebraicity. Second part : We select a maximal ideal I' containing I and we look at the homomorphism

$$k[X_1,\dots,X_n] \longrightarrow k[X_1,\dots,X_n]/I' = k[x_1,\dots,x_n] = K .$$

By construction $x = (x_1,\dots,x_n) \in K^n$ and $P(x) = 0$ for all $P \in I'$, in particular $P(x) = 0$ for all $P \in I \subset I'$.

(1.2) Corollary 1. Let $P_i \in k[X_1,\dots,X_n]$ and K be an extension of k such that $P_i(x) = 0$ for a "point" $x = (x_1,\dots,x_n) \in K^n$. Then there is a point $y = (y_1,\dots,y_n)$ with coordinates in a finite algebraic extension of k such that $P_i(y) = 0$ (for all indices i).

Proof. Let I_K be the ideal of $K[X_1,\dots,X_n]$ generated by I. By hypothesis I_K does not contain 1, hence a fortiori $1 \notin I \subset I_K$. Thus I is strict.

(1.3) Corollary 2. *Let* I *be an ideal of* $k[X_1,\dots,X_n]$ *and* K *an algebraic closure of* k. *Put* $V(I) = V_K(I) \subset K^n$. *Then* $P \in k[X_1,\dots,X_n]$ *vanishes on* V(I) *if and only if a power of* P *is in* I. *In particular if* I *is a prime ideal*, $P \in I$ *if and only if* P *vanishes on* V(I).

Proof (Rabinowitsch). We add an indeterminate X_0 and consider the ideal $\bar{I}$ of $k[X_0,\dots,X_n]$ generated by I and $1 - X_0P$ where P is the polynomial supposed to vanish on V(I). By construction, $\bar{I}$ has no zero on V(I) hence no zero at all. By the Nullstellensatz, $\bar{I}$ is not strict and we can write

$$1 = A_0(1 - X_oP) + \sum A_iP_i \quad ,$$

a finite sum with some polynomials $P_i \in I$. Making $X_0 = 1/P$ in this formal relation leads to a relation

$$\sum A_i(P^{-1},X_1,\dots,X_n)P_i(X_1,\dots,X_n) = 1 \ ,$$

and multiplying by the highest power P^{ℓ} of P appearing in the denominators of the $A_i(P^{-1},X_1,\dots,X_n)$, we get $P^{\ell} \in I$.

Let us fix an algebraic closure K of k. An affine variety in K^n defined over k is by definition a set of the form V(I) for an ideal I of $k[X_1,\dots,X_n]$. This variety is said to be irreducible when the corresponding ideal of $k[X_1,\dots,X_n]$ formed of all polynomials vanishing on it is a prime ideal (by definition, a prime ideal is an ideal whose complement is stable under multiplication and contains 1, or equivalently an ideal giving by quotient an integral domain). By the above corollary 2, irreducible (affine) varieties defined over k and prime ideals of polynomial rings over k correspond one-one to each other. This notion is relative to the field k, and we should say k-irreducible instead of irreducible, to be quite precise. When K is an algebraic closure of k and the ideal I_K generated by I in $K[X_1,\dots,X_n]$ is still prime, we say that $V_K(I) = V_K(I_K) \subset K^n$ is absolutely irreducible (it is irreducible over any field, containing K or not).

As before, let k be any field, and let K be a field extension of k. The set of common zeros of a family (F_i) of homogeneous polynomials (we shall say: forms) of $k[X_o,\dots,X_n]$ in K^{n+1} can be identified with a subset of $\mathbb{P}^n(K)$, the space of lines through the origin in K^{n+1} (or equivalently the quotient of $K^{n+1} - \{0\}$ under the equivalence relation given by homotheties). Again this set of zeros depends only on the ideal I generated by the F_i , and we denote it by

$$V_K = V_K(I) \subset \mathbb{P}^n(K) \quad .$$

This is what we call a projective (sub-)variety of $\mathbb{P}^n(K)$ defined over the field k. We say that an ideal of a polynomial ring is homogeneous when it can be generated by a family of forms. One sees at once that an ideal of polynomials is homogeneous exactly when the homogeneous components of its members are still in it. Thus, a homogeneous ideal has a set of generators constituted by a finite number of forms (and Lefschetz' principle is still valid for projective varieties in characteristic zero). A hypersurface in $\mathbb{P}^n(K)$ defined over k is the set of zeros of a single form F with coefficients in k, or equivalently of a principal homogeneous ideal (F) of $k[X_o,\dots,X_n]$). We shall need (a special case of) a theorem asserting that when K is algebraically closed, a homogeneous ideal with $m \leqslant n$ generators has a zero in $\mathbb{P}^n(K)$ (such an ideal has always a zero in K^{n+1}, namely the origin, but this trivial zero does not correspond to a point of $\mathbb{P}^n(K)$). In other words, the intersection of $m \leqslant n$ hypersurfaces in $\mathbb{P}^n(K)$ is not empty if K is algebraically closed. This is a corollary of the dimension theorem, or of elimination theory, but we prove it directly.

(1.4) <u>Theorem 2</u>. <u>Let</u> k <u>be a field</u>, $1 \leqslant m \leqslant n$ <u>two positive integers, and</u> $(F_i)_{i=1,\dots,m}$ <u>a set of</u> m <u>forms of</u> $k[X_o,\dots,X_n]$. <u>Then there is a finite algebraic extension</u> K <u>of</u> k <u>and a point</u> $0 \neq x = (x_o,\dots,x_n) \in K^{n+1}$

<u>such that</u> $F_i(x) = 0$ <u>for</u> $i = 1,\dots,m$ (<u>if none of the</u> F_i <u>is constant</u>).

<u>Proof</u>. Let $k' = k(F_1(X),\dots,F_m(X)) \subset k(X_o,\dots,X_n)$ with $F_i(X) =$ $= F_i(X_o,\dots,X_n)$. The transcendence degree of k' over k is smaller or equal to m, hence one X_i at least is still transcendent over k'. Let $p = \deg k(X_o,\dots,X_n)/k' \geqslant 1$ and assume that the numbering has been chosen so that $X_1,\dots,X_p$ is a transcendence basis of $k(X_o,\dots,X_n)$ over k'. We put $L = k'(X_1,\dots,X_p)$ and make a diagram.

$k(X_o,\dots,X_n)$
| finite algebraic
$k'(X_1,\dots,X_p) = L$
| purely transcendental of degree $p \geqslant 1$
$k' = k(F_i(X))_{i=1\dots m}$
|
k

We define a discrete valuation on L trivial on k' by $v(X_1) = -1$, $v(X_2) = \dots = v(X_p) = 0$. This defines v uniquely because if P/Q is a representation of an element of L as rational element over k', then $v(P/Q) = \deg_1(Q) - \deg_1(P)$ where $\deg_1$ denotes the degree in X_1 (this is the valuation at infinity associated to X_1). This valuation has an extension to $k(X_o,\dots,X_n)$ which we still denote by v. Let i be such that $v(X_i)$ is minimal, hence $v(X_i) \leqslant -1$. Thus

$$(*)\qquad v(X_j/X_i) = v(X_j) - v(X_i) \geqslant 0 \quad \text{for } j = 0,\dots,n .$$

Now the valuation ring $R_v = \{F \in k(X_o,\dots,X_n) : v(F) \geqslant 0\}$ has a unique maximal ideal $M_v = \{F \in R_v : v(F) > 0\}$, so that $\Omega = R_v/M_v$ is a field containing k' ($\supset$ k). Also

$$(**)\qquad v(F_\ell(X_o/X_i,\dots,X_n/X_i)) = v(F_\ell(X_o,\dots,X_n)/X_i^{d_\ell}) =$$

$$= -d_\ell v(X_i) \geqslant d_\ell = \deg(F_\ell) \geqslant 1 \quad .$$

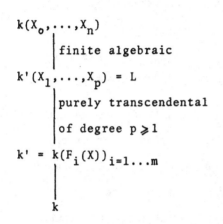

Let y_j be the class of X_j/X_i in Ω (this is meaningful by (*)), so that $0 \neq y = (y_j) \in \Omega^{n+1}$ (remember $y_i = 1$) is a common zero of all F_ℓ by (**). The theorem follows by Cor. 1 of Th. 1 applied to the ideal generated by I and $1 - X_i$ (which has still y as zero). (Alternately, one could also say that if 0 were the only zero of I in K^{n+1}, with an algebraic closure K of k, then I would contain <u>all</u> X_j^N, $j=0,\dots,n$), for a sufficiently large integer N by Cor.2 of Th.1, and thus could not vanish on $0 \neq y \in \Omega^{n+1}$.)

It is convenient to fix an algebraic closure K of k. As in the affine case, we say that a projective variety (in $\mathbb{P}^n(K)$) defined over k is k-<u>irreducible</u>, when its corresponding homogeneous ideal of $k[X_o,\dots,X_n]$ is prime, and we say that it is <u>absolutely irreducible</u> when its corresponding homogeneous ideal of $K[X_o,\dots,X_n]$ is still prime. Let $k[X_o,\dots,X_n] = \bigoplus_{d\geqslant 0} A_d$ be the decomposition of this graded ring in homogeneous components (the A_d are finite dimensional k-vector spaces). A homogeneous ideal I of this graded ring is precisely an ideal I such that $I = \bigoplus I_d = \bigoplus (I \cap A_d)$. By Cor.2 of Th.1, I has 0 as only zero if and only if $I_d = A_d$ for some $d \geqslant 0$ (hence $I_m = A_m$ for all $m \geqslant d$). We shall say that a homogeneous ideal I is <u>strict</u> when $I_d \neq A_d$ for all $d \geqslant 0$ (or equivalently when the corresponding subvariety of $\mathbb{P}^n(K)$ is not empty). Th.2 says that <u>an ideal of</u> $k[X_o,\dots,X_n]$ <u>generated by</u> $m \leqslant n$ <u>non-constant forms is strict</u>. The only prime homogeneous ideal which is not strict is $\bigoplus_{d\geqslant 1} A_d$. Let I be any strict, homogeneous, prime ideal of $A = \bigoplus A_d$, and let V be the corresponding projective variety in $\mathbb{P}^n(K)$. We define the field k(V) of k-<u>rational functions</u> on V to be the field of classes of elements

$$F/G \text{ with } F \text{ and } G \in A_d \text{ for some } d \geqslant 0, \text{ and } G \notin I_d ,$$

with the equivalence relation

$$F_1/G_1 \sim F_2/G_2 \quad \text{whenever } F_1G_2 - F_2G_1 \in I \quad .$$

Because I is a prime ideal, the quotient A/I is an integral domain and

$$k(V) \subsetneq \text{field of fractions of } A/I \quad .$$

All this is done without reference to a k-basis of A_1 (only the gradation of A is used), or as we say, without coordinate system (i.e. without the special linear forms X_i). But if I is as before, one linear form $Y_0 \notin I_1$ and we may even suppose that Y_0 is selected among the forms X_i if we want. After renumbering the indeterminates, we may thus suppose $X_0 \notin I$. We fix now X_0 and classify the strict, homogeneous, prime ideals I which do not contain X_0. With that purpose in mind, we introduce two operations (inverse to each other in a sense), the first one being the homogeneization operation which associates to each polynomial $G \in k[X_1,\ldots,X_n]$ the homogeneous polynomial (form) G^* in the n + 1 indeterminates $X_0,\ldots,X_n$ defined by

(1.5) $$G^*(X_0,\ldots,X_n) = X_0^{\deg(G)} G(X_1/X_0,\ldots,X_n/X_0) \quad .$$

The second one is the dehomogeneization which associates to each form $F \in k[X_0,\ldots,X_n]$ the polynomial in the n indeterminates F_0 defined by

(1.6) $$F_0(X_1,\ldots,X_n) = F(1,X_1,\ldots,X_n) \quad .$$

Making $X_0 = 1$ in the definition of G^*, it is obvious that $(G^*)_0 = G$ for every polynomial G. Moreover it is easily seen that $(F_0)^* = F$ for every form F not divisible by X_0 (from this one can see in general that $(F_0)^* = X_0^{-m} \cdot F$ where m is the largest integer such that F is divisible by X_0^m). These operations are extended to ideals and give a one-one correspondance between (e.g.) strict, prime ideals of $k[X_1,\ldots,X_n]$ and strict, prime, homogeneous ideals of $k[X_0,\ldots,X_n]$ <u>which do not contain</u> X_0 . If we fix again a strict, homogeneous, prime ideal I not containing X_0 , the homomorphism

$$k[X_1,\ldots,X_n] \longrightarrow k(V) \subset \text{field of fractions of } A/I$$

$$G \longmapsto G^*/X_0^{\deg G} \quad ,$$

has for kernel precisely the set of G such that $G^* \in I$ and so consists of those $G = (G^*)_0 \in I_0$. In other words, we have an isomorphism

$$\text{field of fractions of } k[X_1,\ldots,X_n]/I_0 \longrightarrow k(V)$$

given by the preceding map. In particular, we could have defined the field of k-rational functions k(V) on the k-irreducible projective variety V corresponding to the ideal I as being the field of k-rational functions on the k-irreducible affine variety V_0 corresponding to the ideal I_0 (this last field being defined as field of fractions of the integral domain $k[X_1,\ldots,X_n]/I_0$). The drawback of this method is that it depends a priori on the choice of the hyperplane at infinity $X_0 = 0$ (where $X_0 \notin I$).

When we work with a finite number of projective varieties, it is always possible to select a linear form Y_0 (if k is infinite) such that all ideals defining the varieties do not contain Y_0 (note that it may not be possible to choose Y_0 among the X_i here). In other words, it is possible to choose a hyperplane at infinity not containing any member of the finite family. Thus many (local) problems are reduced to affine ones after a suitable choice of hyperplane at infinity.

We also note that by Cor.2 of Th.1, the elements of k(V) have an interpretation as functions defined over non-empty subsets of $V = V_K$, because if F/G is one representation of an element of k(V), $G \notin I$ implies that the set of points of V where G does not vanish is not empty, and on this part of V, F/G defines a function, because F and G are forms of the same degree. Moreover, if F_1/G_1 is another representation of the same element of k(V), it will determine the same function in the intersection of the domains where these functions are defined. Conversely, if F_1/G_1 and F_2/G_2 determine the same function on the subset of V where G_1G_2 does not vanish, they determine the same class in k(V). We do not prove this in general, hence shall refrain from using it (in this general form).

We say that an element $f \in k(V)$ is <u>defined at the point</u> P of V (= V_K and K is an algebraic closure of k) if f admits a representation F/G with either $F(P) \neq 0$ or $G(P) \neq 0$ (in fact F(P) has no meaning because P is a set of points of the form $K^{\times} \cdot P_0$ with $P_0 \in K^{n+1} - \{0\}$ and $F(P) = (K^{\times})^{\deg(F)} F(P_0)$, so that we should write $F(P_0) \neq 0$ or $F(P) \neq \{0\}$). Then f is said to be <u>defined and finite at the point</u> P when moreover, F/G can be selected (in the class of f) so that $G(P) \neq 0$. The set of functions $f \in k(V)$ which are defined and finite at P is a ring R_P. For $f \in R_P \subset k(V)$, the value f(P) is defined unambiguously by $F/G\,(P) = F(P_0)/G(P_0)$ for a $P_0 \in P$. If we assume that P does not lie on the hyperplane of equation $X_0 = 0$ (which we may after renumbering the indeterminates), and if we identify k(V) with the field of fractions of $R = k[X_1,\ldots,X_n]/I_0$, we see that R_P is the subring consisting of fractions F/G with $G(P_0) \neq 0$, where $P_0 = (1,\ldots)$ is in the class of P. The set of $f \in R$ such that $f(P_0) = 0$ is a prime ideal $\mathfrak{p}$ of R (even a maximal ideal if k = K) because the quotient $R/\mathfrak{p} \subset K$. Then R_P is nothing else than the localized of R at $\mathfrak{p}$:

$$R_P = R_{\mathfrak{p}} = R[(\complement\mathfrak{p})^{-1}] \subset \text{field of fractions of } R = k(V) .$$

(1.7) <u>Definition</u>. <u>The subring</u> R_P <u>of</u> k(V) <u>consisting of functions defined and finite at</u> $P \in V_K$ <u>is called local ring of</u> V <u>at</u> P.

This definition is justified by the fact that R_P is a <u>local ring</u> in the sense of commutative algebra

$$R_P - R_P^{\times} = \{f = F/G : F,G \in R \text{ and } F(P) = 0\} = \mathfrak{p}R_{\mathfrak{p}}$$

is an ideal (hence the only maximal ideal of R_P).

(1.8) <u>Proposition</u>. <u>The local ring</u> R_P <u>is a noetherian ring</u>.

Proof. We show that every ideal J_P of R_P is finitely generated. Let $J = J_P \cap R$. Then J is an ideal of the noetherian ring R (a quotient $k[X_1,\dots,X_n]/I_o$ of a noetherian ring is noetherian) and so we can choose a finite family $f_1,\dots,f_N$ of R-generators of J. I claim that the f_i generate J_P (over R_P). Indeed, if $f \in J_P$, we can write $f = g/h$ with $h(P) \neq 0$ and $g = \sum a_i f_i$ $(a_i \in R)$. Then

$$f = \sum (a_i/h) f_i \quad \text{with } a_i/h \in R_P \text{ by definition .}$$

We turn now to the study of curves.

(1.9) Definition. An irreducible variety $V \subset \mathbb{P}^n(K)$ (defined over k) is an algebraic curve (over k) if the transcendence degree of $k(V)$ over k is one.

We shall mainly be interested in irreducible plane curves, i.e. curves in $\mathbb{P}^2(K)$ defined by an irreducible form $F \in k[X_o,X_1,X_2]$. Plane curves are hypersurfaces in $\mathbb{P}^2(K)$, so that we can apply Th.2.

(1.10) Proposition. Let K be an algebraic closure of k and F and G two forms of $k[X_o,X_1,X_2]$ determining two curves $V(F)$, $V(G) \subset \mathbb{P}^2(K)$. If F and G have no common factor, the intersection $V(F) \cap V(G)$ is finite (and not empty).

Proof. We assume that neither F nor G is divisible by X_o, and we prove the assertion of the proposition in the affine piece $X_o = 1$. Thus F_o and G_o have no common factor in $R = k[X_1,X_2]$ (notations of (1.6)) and also no common factor in $k(X_1)[X_2]$ (Gauss' lemma). As this ring is now principal, we can write

$$aF_o + bG_o = 1 \quad \text{with suitable } a,b \in k(X_1)[X_2]$$

and consequently there exists a polynomial $0 \neq C(X_1) \in k[X_1]$ so that

$$AF_o + BG_o = C \neq 0 \text{ with polynomials } A,B \in k[X_1,X_2]$$

The common zeros of F_o and G_o must have a first coordinate among the finite set of roots of the polynomial C. Similarly for the second coordinates of the intersection points (with another polynomial $\mathcal{C}$). This proves that the intersection is finite, because V(F) (and V(G)) have only finitely many points on the line at infinity $X_0 = 0$.

(1.11) Corollary 1. Let $F \in k[X_0, X_1, X_2]$ be a prime form, and V(F) be the corresponding irreducible plane curve. If $f \in k(V)$, the set of points $P \in V(F)$ where f is not defined and finite $f \notin R_P$ is finite.

Proof. Let us select a representation A/B of f, with $B \notin (F)$. Because F is irreducible, B and F have no common factor and we can apply the proposition, because the set of points where f is not defined and finite is contained in $V(B) \cap V(F)$.

(1.12) Definition. Let $F \in k[X_0, X_1, X_2]$ be an irreducible form, and $P \in V(F) \subset \mathbf{P}^2(K)$. The point P is called regular (or simple, or non-singular) on V(F) when the formal derivatives $(\partial F/\partial X_i)(P)$ (i = 0,1,2) do not vanish simultaneously at P (same convention as before regarding the meaning of this). A point which is not regular is called singular. A projective curve is called regular if it has no singular point, and singular if it has at least one singular point. (We shall also say that an affine curve is regular, or non-singular, when the projective curve defined canonically by homogenizing is regular.)

Euler's relation gives

$$\sum X_i(\partial F/\partial X_i) = \deg(F)\cdot F \quad ,$$

and if $P \in V(F)$ is not on the line at infinity $X_0 = 0$, we normalize $P_0 \in P$ by $X_0(P_0) = 1$ (i.e. $P_0 = (1,\cdot,\cdot)$) and we deduce

$$(\partial F/\partial X_0)(P_0) = - X_1(P_0)(\partial F/\partial X_1)(P_0) - X_2(P_0)(\partial F/\partial X_2)(P_0) \ .$$

But obviously $\partial F/\partial X_i = (\partial F_0/\partial X_i)^*$ for i = 1,2 , and so, the conditions

$$(\partial F_0/\partial X_i)(P_0) = (\partial F_0/\partial X_i)^*(P_0) \text{ not both 0 for } i = 1,2$$

are already sufficient for the non-singularity of a point $P \in V(F)$ not on the line at infinity $X_o = 0$.

(1.13) Corollary 2. Suppose that the field k is perfect. Then the set of singular points on an irreducible plane curve defined over k is finite.

Proof. If F and the $\partial F/\partial X_i$ have infinitely many common zeros, they must have a common factor, and hence $\partial F/\partial X_i$ must be a multiple of F (by irreducibility of F). Because the $\partial F/\partial X_i$ are forms of degree $\leqslant \deg(F) - 1$, it follows that they vanish identically, and by Euler's relation

$$\deg(F)\cdot F = \sum X_i(\partial F/\partial X_i) \equiv 0 .$$

This is sufficient in characteristic 0, because $F \equiv 0$ is not a form defining a curve. In characteristic $p \neq 0$, we see that deg(F) must be a multiple of p, and more precisely, the vanishing of the partial derivatives shows that F depends only of the X_i^p. Because k is perfect $F = (F')^p$ by taking the pth roots of all coefficients of F, and so F could not be irreducible. We have obtained a contradiction in both cases.

(1.14) Examples. We consider the following affine curves.

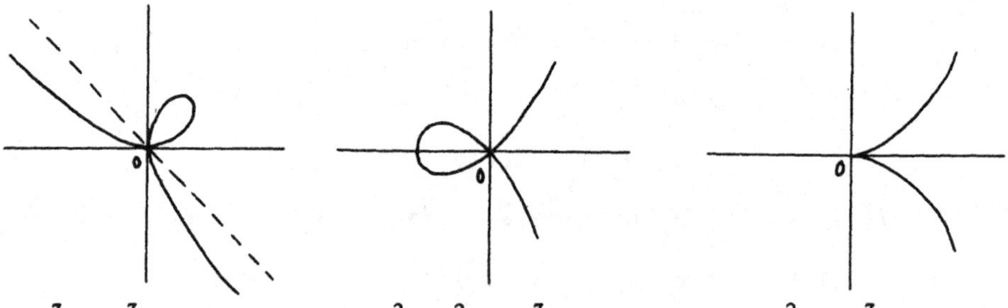

$X^3 + Y^3 - XY = 0$ $\quad$ $Y^2 - X^2 - X^3 = 0$ $\quad$ $Y^2 - X^3 = 0$

They are irreducible, and have all the origin as singular point. The class f of Y/X is not defined at the origin for any of these curves, nor is 1/f defined at the origin for any of them. On the curve of equation $Y - X^3 = 0$, both f and 1/f are defined at the origin, only f being finite at the origin : $f \in R_{(0,0)}(Y-X^3)$.

(1.15) Convention. For the rest of the section, $K = \bar{k}$ will denote a fixed algebraic closure of k, and $V = V_K(F) \subset \mathbb{P}^2(K)$ will be a K-irreducible (i.e. absolutely irreducible) projective plane curve defined over k by a form $F \in k[X_o, X_1, X_2]$.

Let $X = X_K$ denote the set of (discrete) normalized valuations v of K(V) (i.e. such that $vK(V)^* = \mathbb{Z}$) which are trivial on K. This set is going to play the role of "spectrum" of the field K(V) and will be very close to the set of points of V (cf. Cor.1 below). We say that a valuation $v \in X$ is centered at a point $P = P_v \in V$ if for $f \in R_P$

$$v(f) > 0 \text{ is equivalent to } f \in M_P \quad .$$

Obviously v is centered at most at one point P because if $P \neq P' \in V$ there are functions f in $R_P \cap R_{P'}$ with $f \in M_P$ but $f \notin M_{P'}$ (take for f the quotient of two linear forms g/h where h does not vanish at P,P' and g vanishes at P but not at P').

(1.16) Lemma. Every valuation $v \in X$ is centered at a (unique) point $P = P_v \in V$, so that we have a well defined mapping

$$\pi : X \longrightarrow V_K \text{ defined by } v \longmapsto \pi(v) = P_v \quad .$$

Proof. Let X,Y,Z be a permutation of the X_i such that $v(X/Z) \geq 0$ and $v(Y/Z) \geq 0$. Denote by x (resp. y) the class of X/Z (resp. Y/Z) in K(V). If $v(x - \xi) = 0$ for all $\xi \in K$, we would infer that for every polynomial $P(X) \in K[X]$ we would also have $v(P(x)) = 0$, because every polynomial is a product of linear factors in K (algebraically closed). Also $v(R(x)) = 0$ for every rational function $R(X) \in K(X)$ because both numerator and denominator have valuation 0. But as K(V) is algebraic over K(x), it would follow that v is trivial: if $t \in K(V)^*$ write its minimal polynomial $t^N + a_1 t^{N-1} + \ldots = 0$, hence $Nv(t) = v(a_1 t^{N-1} + \ldots)$ which implies $v(t) = 0$. Thus there is a $\xi \in K$ such that $v(x - \xi) > 0$. This ξ is unique, because if ξ' were another element of K with the same property, we would derive $v(\xi' - \xi) = v((x-\xi) - (x-\xi')) \geqslant \mathrm{Inf}(v(x-\xi), v(x-\xi')) > 0$ contrary to the fact that v is assumed trivial on K. Now write the equation of the affine part of V in the form

$$F_o(X,Y) = F_o(\xi,Y) - (X - \xi)A(X,Y) = 0 ,$$

and factor $F_o(\xi,Y) = c\prod(Y - \eta_i)$ in linear terms. We have

$$\sum v(y - \eta_i) = v(x - \xi) + v(A(x,y)) \geqslant v(x - \xi) > 0 .$$

Because x and y have been chosen with positive valuation, every polynomial in x and y will have positive valuation, and the above relation can only hold if for one i at least $v(y - \eta_i) > 0$. But for the same reason as above, this strict inequality can hold at most for one η_i. By construction, the point $P = P_v = (\xi, \eta_i, 1)$ is on the affine curve V_o of equation F_o, and defines a point (still denoted by P) on V. After a translation, we may suppose that this point is at the origin $P_v = (0,0,1)$ of the affine coordinate system chosen. If $f \in R_P$, then f admits a representation A/B with $B(0,0,1) \neq 0$. Because now $v(x) > 0$ and $v(y) > 0$, this implies $v(B_o(x,y)) = 0$ and $v(f) = v(A_o(x,y)/B_o(x,y)) = v(A_o(x,y)) > 0$ is equivalent to $A_o(0,0) = 0$ or to $f \in M_P$ as was to be shown.

We show a kind of converse to this lemma.

(1.17) Theorem 3. Let P be a regular point of V. Then there is a unique valuation $v_P = \mathrm{ord}_P \in X$ which is centered at P. For that valuation, $f \in R_P$ is equivalent to $\mathrm{ord}_P(f) \geqslant 0$. Thus R_P is a valuation ring and $M_P = (\pi_P)$ is principal with generator any element such that $\mathrm{ord}_P(\pi_P) = 1$. Also $\bigcap_{r \geqslant 0} M_P^r = \{0\}$, so that the valuation in question is determined by $\mathrm{ord}_P(\pi_P^r R_P^{\times}) = \mathrm{ord}_P(M_P^r - M_P^{r+1}) = r$. ($K = k$, cf. end of proof.)

If $S = S(V)$ denotes the finite set (K is perfect,(1.13),(1.15)) of singular points of V, we see that $\mathrm{ord} : V - S \longrightarrow X$ defined by $P \longmapsto \mathrm{ord}_P$ is inverse of π ($\pi \circ v$ and $v \circ \pi$ are the identity mapping on the sets where they are defined).

Proof. Let again X,Y,Z be a permutation of the X_i such that $Z(P) \neq 0$ and $(\partial F/\partial Y)(P) \neq 0$. Normalize $P_0 \in P$ by $Z(P_0) = 1$, and identify P_0 with a point (ξ,η) of the affine plane K^2. After a translation (replacing $x - \xi$ by x' and $y - \eta$ by y') we may assume that $P_0 = (0,0)$ is at the origin of the affine coordinate system defined by X,Y. Thus the affine part V_0 of V is defined by the polynomial F_0 with no constant term. Supposing that F_0 is not proportional to Y (trivial case), we may assume

$$(*) \qquad F_0(X,Y) = YA(Y) - B(X,Y)X$$

with $A(0) = 1$ (this is no restriction) and

$$B(X,Y) = B_0(X) + YB_1(X) + \dots$$

with $B_0(X) \neq 0$ (because F_0 is not divisible by Y). We denote also by x and y the classes of X/Z and Y/Z resp. in K(V).

First step. Obviously $M_P = (x,y)$ is generated (over R_P) by x and y. We show (under the preceding hypotheses) that M_P is principal and generated by $\pi_P = x$, or what amounts to the same, that y is a multiple of x in R_P. By (*) we have $yA(y) = B(x,y)x$, hence $y = xB(x,y)/A(y) \in xR_P$ because $A(0) \neq 0$ implies $1/A(y) \in R_P$.

Second step. We show more precisely that there is an integer $r \geqslant 1$ such that $y \in x^r R_P^{\times} = M_P^r - M_P^{r+1}$. This is so with r = 1 if $B(0,0) \neq 0$. If on the contrary $B(0,0) = 0$, we proceed as follows: multiplying by A(y),

$$A(y)B(x,y) = A(y)B_0(x) + B_1(x)A(y)y + \dots \quad ,$$

and writing $B_0(X) = XB_0'(X)$,

$$A(y)B(x,y) = x(B_0'(x)A(y) + B_1(x)B(x,y) \quad + y(\dots)) =$$

$$= x(B_0' + B_1 B_0 + y(\dots)) = x(\widetilde{B}_0 + y(\dots)) \quad ,$$

with a new polynomial $\widetilde{B}_0(X) = B_0'(X) + B_0(X)B_1(X)$ containing exactly one power of X less than $B_0(X)$. If X^{r-1} is the highest power of X dividing B_0, we shall get after r-1 steps

$$A(y)^{r-1}B(x,y) = x^{r-1}(\tilde{\tilde{B}}_0(x) + y(\ldots)) \quad ,$$

with $\tilde{\tilde{B}}_0(0) \neq 0$. This shows that

$$y = x^r C(x,y)/A(y)^r \text{ with } C(0,0) \neq 0 \ ,$$

hence the assertion $(C(x,y)/A(y)^r \in R_P^\times)$.

Third step. We show now that

$$\bigcap_{m\geq 0} M_P^m = \{0\}$$

(this is Krull's theorem valid quite generally for local noetherian rings, but we indicate a direct proof in our particular case). Equivalently we show

$$R_P - \{0\} = \bigcup_{m\geq 0} \pi_P^m R_P^\times = \bigcup_{m\geq 0} (M_P^m - M_P^{m+1})$$

If $f \in R_P - \{0\}$ we choose a representation C/D of f with $D(0,0) \neq 0$. If $C(0,0) \neq 0$, $f \in R_P^\times$ and we are done. If on the contrary $C(0,0) = 0$, it is rather obvious that a method similar to that used in the second step will lead to $f \in x^m R_P^\times$.

The conclusion of the proof is now easy. Any normalized valuation (trivial on K) centered at P is such that $v(x) = 1$ and trivial on $R_P^\times$. This shows that there is at most one such valuation. Conversely we can define for $f = x^m g(x,y)$ with $m \in \mathbb{Z}$ and $g(x,y) \in R_P^\times$ $\mathrm{ord}_P(f) = m$. As every element f of $K(V)^\times$ is of this form this gives the existence of the required valuation, and concludes the proof. We note however, that we have only used the fact that k = K is algebraically closed to be able to suppose that the point under consideration was at the origin. In other words, to have $x - \xi \in k(V)$ and $y - \eta \in k(V)$. Thus, the proof shows that the result is true if we replace K by the subfield $k' = k(\xi,\eta) = k(P)$ generated by the coordinates of P_0 , over any field of definition of V (i.e. containing the coefficients of the polynomial $F_0(X,Y)$ defining V). This field k(P) is a finite algebraic extension of k (independent of the choice of the line at infinity, as long as we take the representant P_0 with one coordinate equal to 1).

(1.18) Corollary 1. If $V = V_K$ is an absolutely irreducible non-singular plane curve, $\pi : X \longrightarrow V$ is a bijection.

(One could show in general that π is surjective and $\pi^{-1}(S(V))$ is finite. In fact X is in one-one correspondence with the normalized - or any non-singular model - of the curve V.)

One should be careful about the fact that even if $\pi^{-1}(P)$ has only one element v, $v(f) > 0$ (for $f \in K(V)$) does not imply $f \in M_P$. Take for example as in (1.14) the curve $y^2 = x^3$, P at the origin, and $f = y/x$. Then there is a unique valuation v centered at P, and for that valuation $v(f) = 1$, but $f \notin M_P$ (in fact $f \notin R_P$).

Recall now that a subring A of a field F is called valuation ring if for any $0 \neq x \in F$, either x or x^{-1} belongs to A. The rings associated to valuations (by $v(f) \geq 0$) as before, are obviously valuation rings in this sense. But conversely

(1.19) Corollary 2. R_P is a valuation ring of K(V) if and only if the point P is regular (and then R_P is the ring of ord_P).

Proof. Remains to show that if R_P is a valuation ring, then P is regular. So we suppose that $y/x \in R_P$ (otherwise interchange y and x). By definition we can write $y/x = A(x,y)/B(x,y)$ with formal polynomials $A(X,Y)$, $B(X,Y) \in K[X,Y]$ and say $B(0,0) = 1$ (we suppose P at the origin). This gives $YB(X,Y) - XA(X,Y) \in I_0 = (F_0)$. A degree consideration shows that up to a constant factor

$$F_0 = YB(X,Y) - XA(X,Y) ,$$

so that $(\partial F_0/\partial Y)(0,0) = 1 \neq 0$ and $P = (0,0)$ is regular.

We note explicitely that all $f \in K(V)$ are defined at all regular points $P \in V$ (possibly infinite at some of them). If f is defined and finite at the regular point P, f(P) is the unique element a_0 of K such that $\mathrm{ord}_P(f - a_0) > 0$. Equivalently we may write $f \equiv a_0 \bmod M_P$. From this it follows that f can be expanded in a formal <u>Taylor series</u>

(1.20) $$T(f) = a_0 + a_1\pi_P + a_2\pi_P^2 + \ldots$$

where a_1 is determined by $a_1 \equiv (f - a_0)\pi_P^{-1} \bmod M_P$, and so on inductively. This series converges to f in R_P (or its completed $\hat{R}_P$) for the topology defined by taking the powers of the maximal ideal M_P as neighbourhoods of 0 (hence all ideals of R_P are neighbourhoods of 0 because R_P is principal). We can also look at this expansion as formal series $T(f) \in K[[\pi_P]]$. More generally, every $f \in K(V)$ can be expanded in a <u>Laurent series</u> at P, which will be an element of the field of fractions $K((\pi_P))$ of the former ring $K[[\pi_P]]$. Then ord_P gives the usual notion of order of a Laurent series.

Suppose now that V is non-singular. Let Div(V) be the free abelian group generated over the set of points of V. Then we can define the <u>divisor</u> of $f \in K(V)$ by the formal expression $\sum \mathrm{ord}_P(f)\cdot(P)$. By (1.11) applied to f and 1/f , only finitely $\mathrm{ord}_P(f)$ will be different from 0 which assures

(1.21) $$\mathrm{div}(f) = \sum_{P \in V} \mathrm{ord}_P(f)\cdot(P) \in \mathrm{Div}(V) .$$

(If V were not assumed to be non-singular, we could define the divisor of f on X using the fact that $\pi^{-1}(S(V))$ is finite, by

$$\mathrm{div}(f) = \sum_{v \in X} v(f)\cdot(v) \in \mathrm{Div}(X) \quad .)$$

Another consequence of Th.3 is the possibility of defining purely algebraically the order of contact of a line and the curve V at a regular point $P \in V$. Let $\alpha X + \beta Y + \gamma = 0$ be the equation of an affine line. We can define the order of contact of the curve and this line at the regular point $P \in V$ to be the positive integer $\mathrm{ord}_P(\alpha x + \beta y + \gamma)$. This order will be 0 exactly when P is not on the line under consideration, it will be $\geqslant 1$ if the line passes through P. When this order is $\geqslant 2$ we say that the line is tangent to the curve at P, and when this order is $\geqslant 3$ we say that this line is an inflexion tangent at P, and that P is a flex of V.

Every regular point has a unique tangent. To see this, we suppose the point at the origin of the affine coordinate system chosen, and we write the defining equation of the curve in the form $F_o(X,Y) = aX + bY + G(X,Y)$ with a polynomial G containing only terms of degree $\geqslant 2$. Then $\mathrm{ord}_P(ax + by) = \mathrm{ord}_P(-G(x,y)) = \mathrm{ord}_P(G(x,y))$ is $\geqslant 2$, so that $aX + bY = 0$ is the equation of a tangent to P (note that $(a,b) \neq (0,0)$ because P is regular). In this case $\pi_P = bx - ay$ is a uniformizing variable (i.e. generator of M_P), and if $a'X + b'Y = 0$ was the equation of another tangent line, we would get $a'x + b'y = \pi_P^2 u$ $(u \in R_P)$ and thus

$$a'X + b'Y - b^2X^2 + 2abXY - a^2Y^2 \equiv 0 \bmod (F_o) \quad ,$$

hence, by degree consideration, the left hand-side would be equal to F_o (up to a constant in K), and this shows (a',b') proportional to (a,b). To find a convenient criterium for the point P to be a flex, we make the further reduction where $a = 0$ (so that $b \neq 0$ and we can take $\pi_P = x$ as uniformizing variable, as in the proof of Th.3), and we write more explicitely the defining polynomial

$$(1.22) \qquad F_o(X,Y) = Y + rX^2 + sXY + tY^2 + G'(X,Y) \quad ,$$

with a polynomial G' containing only terms of degree higher than 3. Then Y = 0 is the equation of the tangent, and

$$\mathrm{ord}_P(y) = \mathrm{ord}_P(rx^2 + sxy + ty^2 + G'(x,y))$$

equals $2\mathrm{ord}_P(x) = 2$ as soon as $r \neq 0$ (because $\mathrm{ord}_P(y) \geqslant 2$). Thus P is a flex precisely when r = 0. We shall transform this condition in a projective form. Let F be the form defining V and let

(1.23) $$H(X_0,X_1,X_2) = \det(\partial^2 F/\partial X_i \partial X_j) \quad .$$

This is a form of degree $\leqslant 3(n - 2)$ if n = deg(F), called the _Hessian_ of F. The plane curve V(H) defined by $H(X_0,X_1,X_2) = 0$ is also called the Hessian of V. If the characteristic of the field K is 0, the Hessian form is of degree 3(n - 2) exactly, and we have the following criterium.

(1.24) _Proposition_. _Suppose that_ K _is of characteristic_ 0, _and let_ V(H) _be the Hessian of_ V. _Then a regular point_ P _of_ V _is a flex precisely when_ $P \in V \cap V(H)$ _is an intersection point of the Hessian (of V) with_ V.

Proof. We take (once more) a permutation X,Y,Z of the X_i such that $Z(P) \neq 0$, $(\partial F/\partial Y)(P) \neq 0$, and assume that P_0 (normalized by $Z(P_0)=1$) is at the origin of the coordinate system defined by X,Y. Moreover we may assume that the tangent line to P is the line Y = 0 and so F_0 will be given by (1.22) of which we take the notations. We compute ZH(X,Y,Z) by multiplying by Z the "last" column of H and we add to this column X times the first one and Y times the second one (this does not change the value of the determinant). Euler's relation in the last column, will give the three terms

$$(n-1)\partial F/\partial X \ , \ (n-1)\partial F/\partial Y \ , \ (n-1)\partial F/\partial Z \ ,$$

because these partial derivatives are forms of degree n-1 (we are in characteristic 0). We do the same trick with the lines, adding

X times the first one, Y times the second one, to Z times the last one. We get

$$Z^2H = \begin{vmatrix} F''_{XX} & F''_{XY} & (n-1)F'_X \\ F''_{YX} & F''_{YY} & (n-1)F'_Y \\ (n-1)F'_X & (n-1)F'_Y & n(n-1)F \end{vmatrix}$$

(with notations for the partial derivatives which are better suited to typewriters!). If $P \in V$ we get thus

$$H(P_o) = \begin{vmatrix} 2r & s & a \\ s & 2t & b \\ a & b & 0 \end{vmatrix} (n-1)^2$$

with the notations of (1.22) where we have supposed $a = 0$, $b = 1$. This shows that the point P_o will be on the Hessian precisely when $r = 0$ in the case of (1.22), which is a necessary and sufficient condition for P to be a flex.

We have used the characteristic assumption to be assured that the partial derivatives are forms of degree n-1 and to conclude $r = 0$ from $2(n-1)^2r = 0$. Thus the characteristic assumption can be relaxed in special cases. For example, if F is a form of degree 3 (defining a cubic V), the result of the proposition will obviously hold in characteristic $p \neq 2, 3$.

(1.25) Corollary. In characteristic 0, a non-singular plane curve of degree > 2 has at least one flex.

Proof. Indeed, the Hessian is of degree $3(n-2) > 0$ in that case, and it is sufficient to apply Th.2 (1.4).

2. Plane cubic curves

Let k be any field, $K = \bar{k}$ an algebraic closure of k. A plane cubic curve defined over k is given by a cubic form

(2.1) $$a_o X_o^3 + (a_1 X_1 + a_2 X_2) X_o^2 + \ldots + a_6 X_1^3 + a_7 X_1^2 X_2 + a_8 X_1 X_2^2 + a_9 X_2^3$$

with ten coefficients $a_i \in k$. Two families (a_i) determine the same cubic curve in $\mathbb{P}^2(K)$ whenever they are proportional (with non-zero factor of proportionality). Thus we can identify cubics in $\mathbb{P}^2(K)$ defined over k with elements of $\mathbb{P}^9(k)$. Some of these cubics are reducible (degenerate in a line plus a conic, or three lines, distinct or not), and some irreducible cubics have singularities. We shall show first, that with mild assumptions, a K-irreducible non-singular such cubic has a very simple equation in a suitable coordinate system. Equivalently, we could say that after a suitable projective transformation of $\mathbb{P}^2(K)$, the curve of equation (2.1) can be brought on a curve with a much simpler equation.

(2.2) Theorem 1. Let V be an irreducible plane cubic defined over the algebraically closed field K of characteristic $p \neq 2$. Suppose that V has a flex P. Then there is a coordinate system in which V has an equation of the form $Y^2Z - F(X,Z) = 0$ with a form F of degree three. Conversely, if F is not divisible by Z, this equation defines an irreducible cubic with a flex on the line $Z = 0$.

Proof. Choose a basis X , Y, Z of linear forms such that the flex has (X,Y,Z) coordinates respectively equal to (0,1,0). Suppose moreover that $Z = 0$ is the tangent to V at the flex P. In the affine coordinate system $Y = 1$, the equation of V will have the form (1.22) with $r = 0$

$$Z + sXZ + tZ^2 + A_3(X,Z) = 0$$

with a homogeneous polynomial A_3 of degree 3. In projective form, this equation will be

$$ZY^2 + sXYZ + tYZ^2 + A_3(X,Z) = 0 .$$

Because we assume that the characteristic is not 2 we may define a new form $Y' = Y + \frac{1}{2}(sX + tZ)$ whence an equation

$$ZY'^2 + A'_3(X,Z) = 0$$

with another homogeneous polynomial A'_3 of degree 3. This is the required form. Conversely, if F is not divisible by Z, the irreducibility amounts to the irreducibility of the polynomial

$$Y^2 - F(X) \text{ in } K[X,Y] \text{ or in } K(X)[Y] \quad \text{(Gauss' lemma)}$$

where F is a polynomial of degree 3. This is obvious because this degree being odd, F cannot be a square. Thus $Y^2Z - F(X,Z) = 0$ is the equation of an irreducible cubic (if F is not divisible by Z), and this cubic has a flex at $(X,Y,Z) = (0,1,0)$.

We emphasize that if V is K-irreducible and non-singular, and if the characteristic p is not 2 or 3, the existence of a flex follows from (1.25) (and the remark just before it), hence

(2.3) Corollary. Every irreducible non-singular cubic in characteristic $p \neq 2,3$ has a Weierstrass equation $Y^2Z = X^3 + aXZ^2 + bZ^3$ with $4a^3 + 27b^2 \neq 0$. An irreducible non-singular cubic in characteristic $p \neq 2$ having a flex, has an equation in Legendre form

$$Y^2Z = X(X - Z)(X - \lambda Z) \qquad \lambda \neq 0 , 1$$

(over the algebraically closed field K).

Proof. The first result follows from the remark above combined with the possibility of completing the cube in characteristic $\neq 3$ and thus killing the factor X^2 in the cubic polynomial in X. The Legendre form follows from a normalization of the three roots of the cubic polynomial.

(2.4) Remark. The Weierstrass form in characteristic $p \neq 2,3$ can be attained over any field of definition k of V containing the coordinates of the flex P. The same remark obviously does not apply to the Legendre form. But we see at once on the Legendre form, that the only irreducible <u>singular</u> cubics having a flex, in characteristic $p \neq 2$ have equations of the form

a) $Y^2Z = X^2(X - Z)$ flex at (0,1,0), double point at (0,0,1) ,

b) $Y^2Z = X^3$ flex at (0,1,0), cusp at (0,0,1) .

Now we turn to the existence of group laws on cubics. Let V be an irreducible plane cubic over the algebraically closed field K of characteristic $p \neq 2$ having a flex P. Let $S = S(V)$ be the set of singular points on V and $R(V) = V - S(V)$ the set of regular points of V. For any two points $Q, Q' \in R(V)$, let Q'' be the third intersection point of V and the line through Q, Q' (this line is the tangent to V if $Q = Q'$ and thus $Q'' = Q$ if $Q = Q' = P$). Then we define $Q \underset{P}{+} Q'$ to be the third point of intersection of V and the line through P and Q'' (the symmetric of Q'' with respect to P). Obviously all intersection points involved are regular on V (look at the two singular cubics above, and note that lines through the singular point of V cut V in only one point, and are not tangent to that point). Not less obvious are

$$Q \underset{P}{+} Q' = Q' \underset{P}{+} Q \quad , \quad Q \underset{P}{+} P = Q \quad ,$$

and if Q^P denotes the symmetric of Q on V (with respect to P),

$$Q \underset{P}{+} Q^P = P \quad .$$

We study this law in the singular cases first. We start with the case of an irreducible cubic with a <u>double point</u> (and $p \neq 2$), hence an equation of the form a) above. The two tangents at the origin (0,0,1) are $Y = X$ and $Y = -X$. In a new system of coordinates admitting these tangents as axis, the equation will have the form $X^3 - Y^3 - XYZ = 0$. We take the affine coordinate system $Y = 1$ which contains all regular points on V, and make a picture.

(2.5)

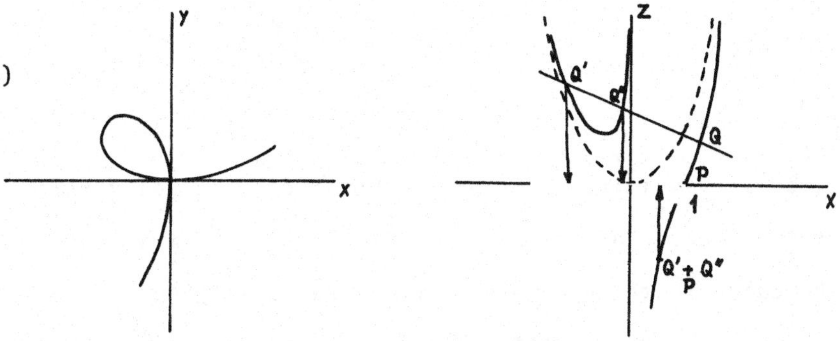

$X^3 - Y^3 - XY = 0$ $(Z = 1)$ $\qquad$ $XZ = X^3 - 1$ $(Y = 1)$

$$z = x^2 - 1/x$$

In the last drawing, we see that a line avoiding the singular point is a non-vertical line of equation $z = mx + h$, and this line cuts the curve in the points of abscissa

$$(mx + h)x = x^3 - 1 \quad , \quad x^3 - mx^2 - hx - 1 = 0 \ ,$$

hence in three points (x_i, z_i) $(i = 1,2,3)$ with $x_1 x_2 x_3 = 1$. This proves that the law $\underset{P}{+}$, where P is the flex $x = 1$, $z = 0$ $(y = 1)$ is isomorphic to the multiplicative group $K^{\times}$ under the first projection $(x,z) \longmapsto x$:

$$\left(R(V), \underset{P}{+}\right) \longrightarrow K \ .$$

This result does not use the fact that K is algebraically closed, and is true as soon as the field k contains the coordinates of the flex and of the double point. The picture shows the case $K = \mathbb{R}$.

Secondly, we take the singular case corresponding to the equation b) above (case of a <u>cusp</u>). Again, we take the affine coordinate system for which $Y = 1$, containing all regular points of V. The flex is then at the origin of this affine system.

(2.6)

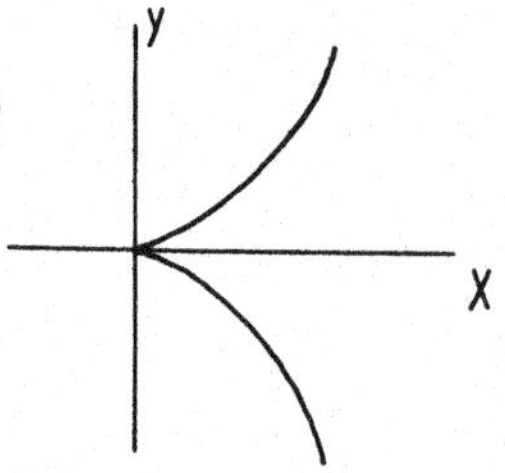

$Y^2 = X^3$ (Z = 1)

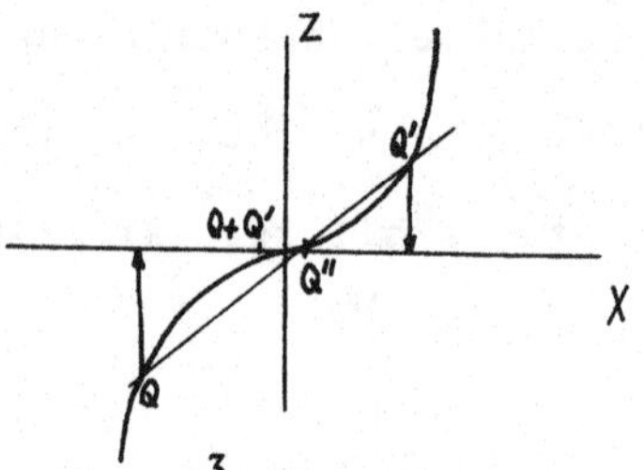

$Z = X^3$ (Y = 1)

As before, a line avoiding the singular point of V is not vertical in this system, hence has an equation of the form $z = m\,x + h$, cutting the curve in three points of abscissa satisfying $x^3 - mx - h = 0$. The three roots of this equation satisfy $x_1 + x_2 + x_3 = 0$ (because the equation does not contain a term in x^2) hence the first projection $(x,z) \longrightarrow x$ gives here an isomorphism

$$R(V), \underset{P}{+} \longrightarrow K^+ \text{ (additive group of K)} ,$$

where P is the flex $x = 0$, $z = 0$ (and $y = 1$). The fact that K is algebraically closed is not more important here than in the last case of the double point, and the picture illustrates the case $K = \mathbb{R}$.

Finally we come to the case where V is <u>non-singular</u> and the characteristic is neither 2 nor 3. With the choice of flex P, we show that V is a group for the law $\underset{P}{+}$. Only remains to check the associativity of this "addition". We choose a coordinate system in which the curve has a Weierstrass' equation, and which puts P on the line at infinity with $Z = 0$ as tangent. Thus we suppose that the affine equation of the curve is $Y^2 = X^3 + aX + b$.
If a line does not go through the chosen flex it is not vertical in this system and has an equation $y = mx + h$, and the intersection points have x-coordinates satisfying $x^3 - m^2x^2 + (a - 2mh)x + b - h^2 = 0$.
If $P_i = (x_i, y_i)$ are these three points, we have

(2.7) $\qquad x_3 = -x_1 - x_2 + ((y_2 - y_1)/(x_2 - x_1))^2 \qquad ,$

where $(y_2 - y_1)/(x_2 - x_1)$ has to be replaced by the slope of the tangent if $x_1 = x_2$ (compare with (I.1.14)). Thus the coordinates of $P_1 \underset{P}{+} P_2$ are rational functions of the coordinates of P_1 and P_2, with coefficients in the prime field. Thus the associativity amounts to the formal verification of the equality of two rational expressions with coefficients in the prime field, or still to the verification that the two polynomials obtained by multiplying the numerator of the first and the denominator of the second on one hand, the denominator of the first and the numerator of the second on the other hand, coïncide. Hence the associativity for the sum of three points (distinct from $P = (0,1,0)$; this is sufficient, because this point acts as unit) is reduced to a formal identity between polynomials. In particular, it is sufficient to check it in characteristic 0 (polynomials with rational integral coefficients). But in this case, it follows from the transcendental theory of chapter I. (Of course we could have avoided this use of Lefschetz principle had we done more algebraic geometry of curves. The interested reader should read the proof given in Fulton, cf. reference given for this chapter.) The group law has been defined in such a way as to give sum 0 (= P) to three points of V on a line. We sum up

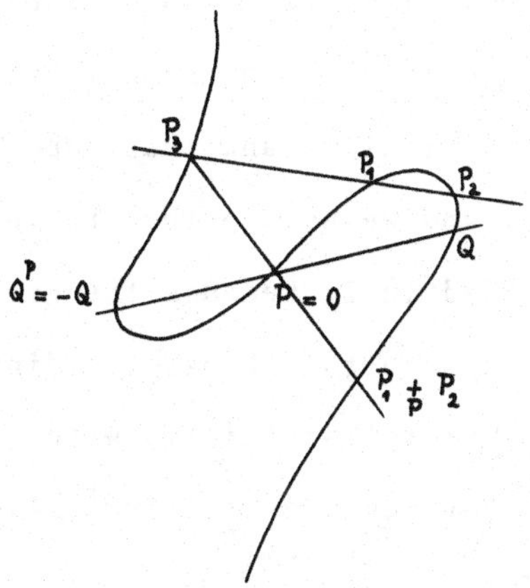

(2.8) Theorem 2. Let V be a non-singular, absolutely irreducible plane cubic defined over the field k of characteristic $p \neq 2,3$, and having a flex P with coordinates in k. Then there is a group law on the set V_k of points of V having coordinates in k, such that P is unit and that three points have zero sum P whenever they lie on a line.

(2.9) Corollary 1. The flexes on a non-singular irreducible plane cubic defined over the algebraically closed field K of characteristic $p \neq 2,3$ (once one of them P is fixed) are the points $Q \in V = V_K$ such that $Q \underset{P}{+} Q \underset{P}{+} Q = P$, or more briefly $3Q = 0$.

Proof. Indeed the condition given is equivalent to $Q \underset{P}{+} Q = Q^P$ because this symmetric Q^P of Q with respect to P is the inverse of Q for the group law deduced from P. This means that the third intersection point of V and the tangent at Q is $Q = (Q^P)^P$ itself.

(2.10) Corollary 2. Let V be an irreducible non-singular plane cubic over the algebraically closed field K of characteristic 0. Then V has nine flexes which make up a subgroup of V (when one of them is chosen as unit element) isomorphic to $(\mathbb{Z}/3\mathbb{Z}) \times (\mathbb{Z}/3\mathbb{Z})$.

Proof. Had we proved Bezout's theorem, the first part would follow immediately, because the Hessian of a cubic is of degree 3 hence has $3 \times 3 = 9$ intersection points with the cubic. We proceed differently (and prove a more precise result: the nine flexes are always distinct). We use Lefschetz' principle and thus suppose that the cubic is defined over $k \subset \mathbb{C}$, and we assume that the algebraic closure $\bar{k}$ of k in K is itself embedded in $\mathbb{C}$. Because the group law is given by rational functions (we suppose that P has coordinates in k, or that the cubic is given in Weierstrass' form) with coefficients in k, the solutions of $3Q = 0$ have coordinates in $\bar{k} \subset \mathbb{C}$. We are reduced to the case $K = \mathbb{C}$, where the situation is easy by the transcendental theory of the first chapter : points of order dividing three on the elliptic curve $\mathbb{C}/L$ are the points $(1/3)L/L \cong L/3L$, which form a subgroup of order 9 isomorphic to $(\mathbb{Z}/3\mathbb{Z})^2$ as was to be proved.

(2.11) Remark. The nine flexes on such a cubic curve have a special configuration. Indeed, if P_1 , P_2 are two distinct flexes, we have $3P_1 = 0$, $3P_2 = 0$ hence $3(-P_1 - P_2) = 0$, so that $-P_1 - P_2$ is a third flex colinear with P_1 and P_2 . To sum up, any line joining two distinct flexes contains a third (distinct) flex. However, the nine flexes are not on a line, because a line cannot cut the cubic in more than three points. In a suitable coordinate system, these nine points will be contained in an affine portion K^2. We cannot make a picture in $\mathbb{R}^2$ of this configuration, because any finite set of points of $\mathbb{R}^2$ such that a line joining two distinct of them contains a third point of the set is necessarily on a line (proof?).

(2.12) Example. Let $y^2 = 4x^3 - g_2x - g_3$ be the (affine) equation of a non-singular plane cubic with g_2 , $g_3 \in \mathbb{Q}$. Then the set of points $(x,y) \in \mathbb{Q}^2$ on this curve, form an abelian group with the point at infinity as neutral element (for the group law described before (2.8)). It can be shown that this abelian group is finitely generated (Mordell-Weil theorem), hence isomorphic to the direct sum of a finite group and a free group $\mathbb{Z}^r$ of rank $r \geqslant 0$. It is still an open problem to determine this rank r in terms of the cubic.

Let k be an algebraically closed field, L <u>a function field of one variable over</u> k, i.e. a (separable) finitely generated extension of k of transcendental degree one. Then it is well known that there exists a transcendental element $x \in L$ such that L is a finite separable algebraic extension of k. Such an extension has a generator y, so that we can write $L = k(x,y)$. Let F be the minimal polynomial of y over k(x) and multiply it by a suitable polynomial in x so as to get an irreducible polynomial $F \in k[X,Y]$. Then, $L = k(V)$ if $V = V(F)$ is the (affine) irreducible variety of zeros of F.
We denote by X the set of normalized valuations of L which are trivial on k (hoping that this X will not be confused with the former indeterminate X !). We have a mapping (1.16)

$$\pi : X \longrightarrow V \quad , \quad v \longmapsto P_v = \pi(v) \quad .$$

This mapping is bijective outside the set S(V) of singular points of V. We <u>assume</u> that $\pi^{-1}(S(V))$ is <u>finite</u> (this could be proved, but we know at least that it is true if V is non-singular). Thus every $f \in L$ has a well defined divisor

$$\mathrm{div}(f) = \sum_{v \in X} v(f) \cdot (v) \in \mathrm{Div}(X) \quad .$$

For every divisor $\underline{d} \in \mathrm{Div}(X)$ we define the k-subspace $L(\underline{d})$ of L by

$$L(\underline{d}) = \{ f \in L : \mathrm{div}(f) \geqslant -\underline{d} \}$$

(remember the convention $\mathrm{div}(0) \geqslant -\underline{d}$ for any divisor $\underline{d}$, so that $0 \in L(\underline{d})$).

(2.13) <u>Definition</u>. <u>We say that</u> L <u>is of genus</u> g <u>over</u> k <u>when the preceding conditions are satisfied and when</u>

$$\dim_k L(\underline{d}) = \deg(\underline{d}) + 1 - g \quad \text{for all } \underline{d} \in \mathrm{Div}(X) \text{ of } \deg(\underline{d}) > 2g - 2.$$

Observe that $h \longmapsto fh$ defines a k-linear isomorphism of $L(\underline{d})$ onto $L(\underline{d} - \mathrm{div}(f))$. In particular, these spaces have the same dimension and this proves that the degree of the divisors of the form div(f) are always zero in a field of genus g.

(2.14) Example 1. The rational field $k(x)$ is of genus 0. Indeed, in this case, the set X is in one to one correspondence with the projective line over k, and divisors $\underline{d}$ over X are thus expressions of the form $\underline{d} = d_\infty(\infty) + \sum_{a\in k} d_a(a)$ with rational integers d_a vanishing for all but a finite number of $a \in k$. To compute the dimension of $L(\underline{d})$, we observe that $f \longmapsto f \prod_{a\in k}(x-a)^{d_a}$ is a k-linear isomorphism of $L(\underline{d})$ into the vector space of polynomials $k[x]$, having as image the subspace of polynomials F such that

$$-\deg(F) = \mathrm{ord}_\infty(F) \geqslant -d_\infty - \sum_{a\in k} d_a = -\deg(\underline{d}) \ ,$$

or equivalently, such that $\deg(F) \leqslant d = \deg(\underline{d})$. This subspace of polynomials is of dimension $d+1$, so that the conditions of the definition (2.13) are satisfied with $g = 0$.

(2.15) Example 2. The function field $k(V)$ over a singular plane cubic curve V is of genus 0. It is sufficient to look at the two special curves defined in (2.4). If the singular cubic has an equation $y^2 = x^3$, it can be parametrized by rational functions, e.g. $y = t^3$ and $x = t^2$. If the singular cubic has an equation $y^2 = x^2(x-1)$, it can also be parametrized by rational functions, e.g. $x = 1 + t^2$, $y = t(1+t^2)$. In both cases, $k(V) = k(x,y) \subset k(t)$ is a rational field (Lüroth's theorem). It has genus 0 by the example 1 above.

(2.16) Example 3. We have proved in (I.2.12) that if L is a lattice in $\mathbb{C}$, then the field of L-elliptic functions is a field of genus one over the complex field.

The above examples exhaust the possibilities for $g = 0,1$ in the following more precise sense.

(2.17) Theorem 3. *Let k be an algebraically closed field, L a function field of one variable over k. If L is of genus 0 over k, then L is purely transcendental L = k(x) (L is a rational field). If L is of genus one over k, then L is the field of rational functions over a non-singular irreducible plane cubic with a flex. In this case (g = 1), if the characteristic p of k is different from 2, the cubic can be taken in Legendre's form* $y^2 = x(x-1)(x-\lambda)$, *and if the characteristic p of k is different from 2 and 3, the cubic can be taken in Weierstrass' form* $y^2 = x^3 + ax + b$. *In these forms, the non-singularity amounts to* $\lambda \neq 0,1$ *and respectively* $4a^3 + 27b^2 \neq 0$.

Proof. Observe once for all that X is not empty : L contains a rational subfield and all valuations from this rational subfield (trivial on k) can be extended to L (this shows more precisely that X is infinite because k is infinite).

First part. Assume $g = 0$, and take any $v \in X$, putting $\underline{d} = (v)$ for the corresponding divisor on X of degree one. By hypothesis $L(\underline{0})$ is of dimension 1, hence equal to k, and contained in $L(\underline{d})$ of dimension 2. Select any $x \in L(\underline{d}) - L(\underline{0}) \subset L$, so that necessarily $L(\underline{d}) = k \oplus xk$. By construction $v(x) = -1$ because $x \notin L(\underline{0})$, and consequently $v(x^n) = -n$ for all integers n. By induction on n, one sees that $L(n\underline{d}) = k \oplus xk \oplus \ldots \oplus x^n k$. I claim that the rational subfield k(x) of L is *the whole field* L. Take any f in L - k and write its divisor $\mathrm{div}(f) = (f)_0 - (f)_\infty$ (symbolically $f^{-1}(0) - f^{-1}(\infty)$), where $(f)_\infty \geqslant 0$ is the polar part of the divisor of f. In particular $\mathrm{div}(f) \geqslant -(f)_\infty$ and so $f \in L((f)_\infty)$. This shows that the $2(n+1)$

functions

$$1\ ,\ x\ ,\ x^2\ ,\ \dots\ ,\ x^n\ ,$$
$$f\ ,\ xf,\ x^2f,\ \dots\ ,\ x^nf\ ,$$

are in $L(n\underline{d} + (f)_\infty)$. By hypothesis this last space is of dimension $n + d_\infty + 1$ where $d_\infty = \deg(f)_\infty$. As soon as $n \geqslant d_\infty$, there is a strict inequality $n + d_\infty + 1 < 2(n+1)$. Hence there must be a linear dependance relation over k between the former functions, and this leads to an equation

$$\sum a_i x^i \cdot f = \sum b_i x^i \qquad (a_i\ ,\ b_i \in k,\ 0 \leqslant i \leqslant d_\infty).$$

As the x^i are linearly independant over k , the coefficients a_i cannot vanish simultaneously. This proves $f = (\sum b_i x^i)/(\sum a_i x^i)$ is in $k(x)$ as was to be proved.

<u>Second part</u>. Assume now $g = 1$ and take $\underline{d} = (v)$ with $v \in X$ as before. Because $L(\underline{d})$ is of dimension 1 now, we must have $L(\underline{d}) = k$. So we select $x \in L(2\underline{d}) - L(\underline{d})$. This element must have $v(x) = -2$. Since $L(3\underline{d})$ is of dimension 3 over k we can select $y \in L$ such that $L(3\underline{d}) = k \oplus xk \oplus yk$, and this implies $v(y) = -3$. From this we derive

$$L(4\underline{d}) = k \oplus xk \oplus yk \oplus x^2k \qquad (\text{because } v(x^2) = -4)\ ,$$
$$L(5\underline{d}) = k \oplus xk \oplus yk \oplus x^2k \oplus xyk \qquad (v(xy) = -5)\ ,$$
$$L(6\underline{d}) = k \oplus xk \oplus yk \oplus x^2k \oplus xyk \oplus x^3k$$
$$= k \oplus xk \oplus yk \oplus x^2k \oplus xyk \oplus y^2k \quad ,$$

because both $v(x^3) = v(y^2) = -6$. This shows the existence of a linear relation over k between the powers in question, that is, a cubic relation between x and y of the form

$$(2.18) \qquad F(x,y) = y^2 + (\alpha x + \beta)y + \gamma x^3 + \delta x^2 + \varepsilon x + \eta = 0$$

with coefficients in k. If the characteristic is not two, we can replace $y + \frac{1}{2}(\alpha x + \beta)$ by y and thus kill the term in y, getting the equation $y^2 = \gamma x^3 + \tilde{\delta} x^2 + \tilde{\varepsilon} x + \tilde{\eta}$. The Legendre form is obtained from there by normalizing x and y in order to get a unitary polynomial in x, and by normalizing its roots as indicated. We have repeatedly said that then $\lambda \neq 0,1$ is the non-singularity condition, a fact which can be checked in any characteristic $p \neq 2$ (the only possible singular points occur with ordinate $y = 0$, and their abscissa must be simultaneous roots of the cubic polynomial in x and its derivative). In characteristic $p \neq 3$, the possibility of killing the term in x^2 is open by completing the cube (and non-singularity condition as claimed if $p \neq 2,3$). Remains to show that the subfield $k(x,y)$ of L is equal to L. As in the first part, we take any $f \in L - k$, and consider the $3(n+1)$ functions

$$\begin{array}{l} 1 \, , x \, , x^2 \, , \dots \, , x^n \quad , \\ y \, , xy, x^2y, \dots \, , x^ny \quad , \\ f \, , xf, x^2f, \dots \, , x^nf \quad . \end{array}$$

They are all in $L((2n+3)\underline{d} + (f)_\infty)$ of dimension $2n+3+d_\infty$ (notations as in the first part). As soon as n is sufficiently large ($n \geqslant d_\infty + 1$ will do), they will be linearly dependant over k, hence a relation

$$\sum a_i x^i + y \sum b_i x^i = f \sum c_i x^i \quad ,$$

with a non-zero polynomial $\sum c_i x^i$ (otherwise, $y \in k(x)$: a sheer nonsense in view of the fact that $v(x)$ is even and $v(y)$ odd). This gives eventually $f = r(x) + s(x)y \in k(x,y)$ with $r(x)$, $s(x) \in k(x)$ (and gives once more the fact that y is quadratic over $k(x)$ by taking $f = y$). Let us just recall that the non-singularity of the curve follows from example 2 above. q.e.d.

No doubt the reader will have recognized the analogy between the two functions x and y over the curve and the Weierstrass' functions $\wp$ and $\wp'$ (the transcendental possibility of taking for y the derivative of x calls for the factor 4 in $y^2 = 4x^3 - g_2x - g_3$).

With the notations of the proof (case g = 1), the divisor $3\underline{d}$ is called very ample, because $L(3\underline{d})$ contains a set of generators of L over k. The divisor $\underline{d}$ itself is called ample because one of its multiples is very ample. This terminology however, adds little to the understanding of the present simple situation.

We shall eventually show that conversely, all function fields over non-singular cubics are of genus one. For that, we have to give a purely algebraic proof replacing that of the first chapter using theta functions. The first steps consist in finding substitutes for Liouville's theorem and Abel's condition (I.1.1a,c,d). Also, the whole idea behind the use of the subspaces $L(\underline{d})$ of a function field L is that they give quite generally a filtration of the k-vector space L with finite dimensional subspaces ($\underline{d} \leqslant \underline{d}'$ implies $L(\underline{d}) \subset L(\underline{d}')$: this is why $L(\underline{d})$ is defined by $\mathrm{div}(f) \geqslant -\underline{d}$ with a minus sign !).

(2.18) Proposition. Let L be a function field over k.

a) $L(\underline{0}) = k$, or in other words : a function $f \in L$ without pole is constant, $f \in k$.

b) The spaces $L(\underline{d})$ are finite dimensional for all divisors $\underline{d} \in \mathrm{Div}(X)$, and more precisely, if $\underline{d} \geqslant 0$ is a positive divisor of degree d $\dim_k L(\underline{d}) \leqslant d + 1$.

c) If $\dim_k L(\underline{d}) = \deg(\underline{d}) + 1$ for one positive divisor $\underline{d} \neq 0$, then L is rational (of genus 0 over k).

Proof. a) If $f \notin k$, define a valuation of k(f) by $v(f) = -1$. If w denotes a normalized extension of v to L, we have $w(f) < 0$ so that f has at least the pole $w \in X$. The same argument applied to $1/(f - a)$ shows that a non-constant function $f \in L$ takes all values $a \in k$.

b) We prove the more precise assertion concerning positive divisors.

Let $\underline{d} \geqslant 0$ and $v \in X$, and let us show that $\dim L(\underline{d} + (v)) \leqslant \dim L(\underline{d}) + 1$. We suppose that $\pi(v) = P_v \in V$ is a non-singular point (cf. note after proof) and let π_P be a uniformizing variable at $P = P_v$: $(\pi_P) = M_P \subset R_P$, that is $v(\pi_P) = 1$. We consider the linear map $L(\underline{d} + (P)) \longrightarrow k$ defined by $f \longmapsto (\pi_P^{d_P+1} f)(P)$ where d_P is the multiplicity of P in $\underline{d}$. By definition its kernel is precisely $L(\underline{d})$, and this proves the desired inequality. Using a), induction applies and proves b).

c) If $\dim L(\underline{d}) = d + 1$ for one positive divisor $\underline{d} \neq 0$ of degree d, then $\dim L((P)) = 2$ by b) for all points P occuring in $\underline{d}$ (with multiplicity $d_P > 0$). Hence for one such point P

$$L((P)) = k \oplus xk \ , \quad \mathrm{ord}_P(x) = -1 \ .$$

Consequently $L(n(P))$ is of maximal dimension $n + 1$ for all positive integers n. Take any $y \in L$, and multiply it by a suitable polynomial in x in order to make it integral over $k[x]$. Then y will have poles only where x has poles, and $y \in L(N(P))$ for a suitably big integer N. This shows $y \in k(x)$ and $L = k(x)$ is rational.

It is true that we have performed the proof in the non-singular case. More precisely, we have used only the following fact : if $v \in X$, there exists a plane curve V such that $L = k(V)$ (a model of L) and that v is centered at a non-singular point $P_v \in V$. However, we shall only use the results of the proposition in the non-singular case.

(2.19) Corollary. If the irreducible plane curve V admits one rational function $f \in k(V)$ with only one pole, and a simple one, at the regular point $P \in V$, then V is isomorphic to a projective line: $k(V) = k(f)$ is a rational field (of genus 0 over k).

This is a particular case of part c) of the above proposition. In particular, f has only one zero and $\mathrm{div}(f) = (P_1) - (P_2)$.

(Some readers will probably have noted the analogy between this corollary and the characterization of the sphere $S^2 = \mathbb{P}^1(\mathbb{C})$ - e.g. among real compact surfaces - by the existence of a Morse function with only two critical points.)

We come back to the case of cubics for Abel's condition (analogous to (I.1.1.d)). Let thus V be an irreducible non-singular plane cubic over the (algebraically closed) field k of characteristic $p \neq 2$. We suppose that the coordinate system has been chosen in such a way that V has the (affine) equation

(2.20) $$V_0 : \quad Y^2 - (X-e_1)(X-e_2)(X-e_3) = 0$$

with distinct $e_i \in k$. We could normalize the e_i's so as to have $e_1 = 0$, $e_2 = 1$, $e_3 = \lambda \neq 0,1$ and work with this equation in Legendre form. However, we prefer to keep more freedom, because if moreover $p \neq 3$, we might prefer to work with the Weierstrass' form (corresponding to the other normalization $e_1 + e_2 + e_3 = 0$).

(2.21) Lemma 1. *If we denote by* ord_∞ *the normalized valuation corresponding to the flex* P *at infinity, we have* $\mathrm{ord}_\infty(x) = -2$ *and* $\mathrm{ord}_\infty(y) = -3$.

Proof. We have to remember that x is the class of X/Z (y that of Y/Z) in the function field L = k(V) in our projective definition of this field. By construction of the coordinate system, P is at the origin of the affine coordinate system defined by Y = 1 and the line Z = 0 is tangent at the flex P, so that

$$\mathrm{ord}_P(\text{class of } Z/Y) = 3 \ , \ \mathrm{ord}_P(\text{class of } X/Y) = 1 \ ,$$

because the line of equation X = 0 goes through P but is not tangent (distinct from Z = 0). This shows

$$\mathrm{ord}_\infty(y) = \mathrm{ord}_P(\text{class of } Y/Z) = -3 \quad \text{as asserted,}$$

and $\quad 1 = \mathrm{ord}_P(\text{class of } X/Y) = \mathrm{ord}_\infty(x) - \mathrm{ord}_\infty(y) = \mathrm{ord}_\infty(x) + 3$.

(2.22) Lemma 2. $L = k(V)$ is not of genus 0.

Proof. The automorphism σ of the quadratic extension L of $k(x)$ defined by $y \longmapsto -y$ acts on the set of normalized valuations X of L and also on the set of points $Q \in V$. Obviously Q^σ is the point with coordinates $x(Q),-y(Q)$ if $Q \neq P$ and $P^\sigma = P$ is fixed under this action. Let us take now a function $f \in L$ which is regular except at P (f is defined and finite at all $Q \neq P$). Then f^σ will have the same property and if $f = a(x) + b(x)y$, $f^\sigma = a(x) - b(x)y$. We show that $f \in k[x,y] = k[V_0]$ the ring of regular functions on the affine part V_0 of V. It is sufficient to show that $a(x)$ and $b(x) \in k[x]$. We prove it for example for a, by decomposing a in linear terms (k is algebraically closed) : $a(x) = \prod_\xi (x - \xi)^{n_\xi}$. If one integer n_ξ were strictly negative, a would have poles at the intersection points of V_0 with the vertical $X = \xi$. We apply this to the determination of $L((P)) \subset k[x,y]$. But a non-constant polynomial in x has order at infinity $\leqslant -2$ (by lemma 1) and a polynomial in x and y containing y has order at infinity $\leqslant -3$. This proves $L((P)) = k$ is of dimension 1 so L cannot be of genus 0.

(2.23) Lemma 3. For $Q \in V_0$ (i.e. $P \neq Q \in V$) and $m,h \in k$, we have

$$\mathrm{div}(x - x(Q)) = -2(P) + (Q) + (-Q) \qquad ,$$

$$\mathrm{div}(y - mx - h) = -3(P) + (Q_1) + (Q_2) + (Q_3) \qquad ,$$

where the points $Q_i \in V$ are the intersection points of V_0 with the line of equation $Y - mX - h = 0$.

Proof. We have already determined the coefficients of (P) in these divisors in the first lemma (the pole of order 2 of x at P cannot compensate the pole of order 3 of y at the same point). It is obvious that $x - x(Q)$ vanishes only at the two points Q and $Q^\sigma = Q^P = -Q = (x(Q),-y(Q))$. The multiplicities of these zeros are given by the order of contact of the vertical $X - x(Q) = 0$ and V at Q.

Because $(\partial/\partial Y)(Y^2 - (X-e_1)(X-e_2)(X-e_3))$ can vanish at Q only if $y(Q) = 0$, we find that the vertical $X - x(Q) = 0$ is not tangent to V at Q (hence the multiplicity is 1) unless $y(Q) = 0$, i.e. $Q = -Q$. In this last case the multiplicity is 2 because these points are not flexes on V. The first formula is thus completely proved. The second one is derived similarly.

We note that lemma 3 gives the divisors of all linear functions on V and that their degrees vanish. Also, they satisfy Abel's condition (when $p \neq 3$ so as to have the group law $\underset{P}{+}$ on V). We give explicitely the special case $m = h = 0$ with a consequence, in the following lemma.

(2.24) Lemma 4. Let $E_i (i = 1,2,3)$ be the three intersection points of V_0 with the X-axis of equation $Y = 0$, so that $E_i = (e_i, 0)$. Then

$$\mathrm{div}(x - e_i) = -2(P) + 2(E_i) \quad ,$$

$$\mathrm{div}(y) = -3(P) + (E_1) + (E_2) + (E_3) \quad .$$

The subgroup of V (when moreover $p \neq 2,3$) of points Q such that $2Q = 0$ is $\{E_0 = P = 0, E_1, E_2, E_3\}$ isomorphic to $\mathbb{Z}/2\mathbb{Z} \times \mathbb{Z}/2\mathbb{Z}$.

The assumption that the characteristic p is not 3 is not necessary, as we shall presently see. But in characteristic $p = 2$, we shall see much later (in the third chapter) that there can at most be one non-zero point of order two, and that the case where $E_0 = P$ is the only point of order dividing 2 also occurs (supersingular curves).

Now we come to a closer study of the subgroup P(X) of principal divisors (of the form div(f) for some $f \in L^\times$) of Div(X). If $\underline{d} = \sum d_Q(Q) \in \mathrm{Div}(X)$ is any divisor, we put $d = \deg(\underline{d}) = \sum d_Q$ and $Q_{\underline{d}} = \sum d_Q \cdot Q$ (sum in the group V with respect to $\underset{P}{+}$). Then the principal divisors $\underline{d}$ are exactly those for which $d = 0$ and $Q_{\underline{d}} = 0$:

(2.25) Theorem 4 (Abel). The group of principal divisors on X is a subgroup of $\mathrm{Div}^{o}(X)$, the group of divisors of degree 0. Moreover,

$$\Phi : V \longrightarrow \mathrm{Div}^{o}(X)/P(X) \quad \text{is an isomorphism}$$
$$Q \longmapsto (Q) - (P) \bmod P(X) \qquad .$$

In particular, the group law on V is independent - up to isomorphism - of the choice of the flex P as neutral element.

(As the law $\underset{P}{+}$ has been shown associative only if $p \neq 2,3$, we have to make that assumption in the above formulation. However, a more careful analysis shows that we only use the two properties $Q \underset{P}{+} Q^{P} = P$ and $(Q_1 \underset{P}{+} Q_2) \underset{P}{+} Q_3 = Q_1 \underset{P}{+} (Q_2 \underset{P}{+} Q_3) = P$ whenever the Q_i's are on a line. Thus this theorem gives another proof for the associativity, which works for the characteristics $p \neq 2$.)

Proof. All linear functions in $k[x,y]$ have principal divisors in $\mathrm{Div}^{o}(X)$ by lemma 3, and they satisfy also the condition $Q_{\underline{d}} = 0$. Take now any principal divisor $\mathrm{div}(f) = \sum d_Q(Q) \in P(X)$. Modulo divisors of linear functions, we have

$$\sum d_Q(Q) \quad = \quad (d-1)(P) + (Q_{\underline{d}}) \ ,$$

so that it is sufficient to prove $d = 0$ and $Q_{\underline{d}} = 0$ for divisors of the form $\mathrm{div}(f) = (d-1)(P) + (Q_{\underline{d}})$. But $Q_{\underline{d}} \neq P$ implies that f has only one simple zero at $Q_{\underline{d}}$ (because f having a pole, $d-1<0$) which is impossible by (2.19) applied to $1/f$, L not being of genus 0 (lemma 2). Then $\mathrm{div}(f) = d(P)$ is possible only with $d = 0$ (loc. cit.) and with $f \in k$. Now we may consider $\mathrm{Div}^{o}(X)/P(X)$ and define Φ as indicated. This map is a homomorphism in virtue of lemma 3 which shows that

$$\Phi(-Q) = -\Phi(Q) \qquad ,$$
$$\Phi(Q_1) + \Phi(Q_2) + \Phi(Q_3) = 0 \quad \text{for } Q_1 + Q_2 + Q_3 = 0 \ .$$

It is injective because $\Phi(Q_1) = \Phi(Q_2)$ implies that $(Q_2) - (Q_1)$ is a principal divisor, and this can happen as above (L is not of genus 0) only when $Q_1 = Q_2$. Let us commit the crime of "lèse Bourbaki" and show that Φ is <u>surjective</u>. If $\underline{d}$ is a divisor of degree zero,

$$\underline{d} = \sum d_Q(Q) = \sum d_Q((Q) - (P)) \quad ,$$

and this gives

$$\underline{d} \bmod P(X) = \sum d_Q \, \Phi(Q) = \Phi(\sum d_Q Q) \quad \text{as in (I.1.24).}$$

The theorem is completely proved, and we are able to turn to the converse of the theorem 3 and show that cubics, when non-singular, have function fields of genus 1.

(2.26) <u>Theorem 5 (Riemann-Roch)</u>. <u>Let</u> V <u>be an irreducible non-singular plane cubic over the (algebraically closed) field</u> k <u>of characteristic</u> $p \neq 2$. <u>Then its rational function field</u> $L = k(V)$ <u>is of genus</u> 1.

<u>Proof</u>. We start by showing that the space $L(n(P))$, where n is a strictly positive integer, is of dimension n. We know that it is contained in $k[x,y]$ (proof of lemma 2). The functions x^i, $x^j y$ for $0 \leqslant i \leqslant n/2$ and $2j + 3 \leqslant n$, i.e. $0 \leqslant j \leqslant (n-3)/2$ make up a basis of $L(n(P))$ because they are linearly independent (functions of the form x^i have a pole of even order 2i at P and $x^j y$ have a pole of odd order at P), and there are exactly n such functions (only a function with a simple pole at P is missing). Then we take an arbitrary divisor $\underline{d} = \sum d_Q(Q)$ of degree $d > 0$. We have already seen that $\underline{d} - d(P) = (Q_{\underline{d}}) - (P) + \mathrm{div}(f)$ with a suitable $f \in L^\times$ (Th.4). Hence $L(\underline{d}) = L((d-1)(P) + (Q) + \mathrm{div}(f)) \cong L((d-1)(P) + (Q))$, the last isomorphism being given by multiplication : $h \longmapsto h\cdot f$. Remains to show that this last space is of dimension d, or that it contains one function not in $L((d-1)(P))$ (cf.(2.18.b) proof) for $d>1$.

We have to construct a function in $L((P) + (Q))$, not constant. We observe that if $Q \neq -Q$, the divisor $-(Q) - (P) + (-Q) + (2Q)$ is principal by Th.4, so the function of which it is the divisor will do in this case. If $Q = -Q$, then Q is one E_i $(i=1,2,3)$. We suppose $Q = E_1$ (for example), and replace the latter divisor by $-(P) - (E_1) + (E_2) + (E_3)$ which is also principal (by the same criterium of Th.4 or more directly using lemma 4, we see that it is the divisor of $y/(x - e_1)$). This concludes the proof.

Let us turn to a full classification of function fields of one variable over k (algebraically closed and of characteristic $p \neq 2,3$) of genus 1. As before, let X be the set of normalized valuations of L trivial on k, and fix one $v \in X$. For a choice of generators x,y of L such that

(2.27) $\quad x \in L(2(v)) - k \ , \ y \in L(3(v)) - L(2(v)) \ ,$

$$y^2 = x^3 + ax + b \quad ,$$

we define

(2.28) $\quad j_v = j_v(x,y) = 1728\,(4a^3)/(4a^3 + 27b^2) \in k \quad .$

This constant is well defined, independently of the choice of the generators x,y for if x',y' are other generators such as in (2.27), and satisfy $y'^2 = x'^3 + a'x' + b'$, we have

$$x' = \alpha x + \beta \quad , \ y' = \gamma y + \delta x + \varepsilon \quad \text{with } \alpha\gamma \neq 0, \beta, \delta, \varepsilon \in k \ .$$

Substituting, we find immediately $\delta = 0$ (no cross term xy in (2.27)) $\varepsilon = 0$ (no term in y) $\beta = 0$ (no term in x^2), so that finally $x' = \alpha x$, $y' = \gamma y$ and $\gamma^2 = \alpha^3 \in k$. If we put $t = \gamma/\alpha \in k$, then $t^2 = \alpha$, $t^3 = \gamma$ and $x' = t^2 x$, $y' = t^3 y$. This shows that the invariant j'_v deduced from the choice of the generators x',y' is equal to j_v . More important is the observation that j_v is actually independent of the choice of $v \in X$.

(2.29) Lemma 5. There is a field automorphism $\tau = \tau(v',v)$ of L over k which brings any $v \in X$ on any (other) $v' \in X$: $\tau L(n(v)) = L(n(v'))$ for all positive integers n.

Proof. Fix $v \in X$, and let V be the non-singular cubic defined by a choice of generators x,y of L with $v(x) = -2$, $v(y) = -3$ as before. Then $\pi(v) = P_v$ is the flex at infinity on V. We consider the translation mappings

$$X \longrightarrow X \quad \text{defined by} \quad w \longmapsto w - v' \quad ,$$

$$V \longrightarrow V \quad \text{defined by} \quad Q \longmapsto Q - Q' \quad (Q' = P_{v'})$$

defined by the group law $+ = \underset{P}{+}$ on V. Because the addition is defined rationally, the coordinates of $Q - Q'$ are rational expressions of the coordinates of Q (with coefficients in k depending on Q').

Call $x^\tau = r_1(x,y)$, $y^\tau = r_2(x,y)$ these two rational expressions :

$$x^\tau(Q) = r_1(x,y)(Q) = r_1(x(Q),y(Q)) = x(Q - Q') \quad ,$$

and similarly $y^\tau(Q) = y(Q - Q')$. Now the mapping

(2.30) $$f = f(x,y) \longmapsto f^\tau = f(x^\tau,y^\tau) : L \longrightarrow L \quad ,$$

is a k-isomorphism (it is surjective because its inverse is given by the consideration of the inverse translation $Q \longmapsto Q + Q'$). (Note that every $f \in L$ has a unique representation $f = a(x) + b(x)y$.)

By definition, $f^\tau(Q + Q') = f(Q)$. In particular

$$L(n(P))^\tau = L(n(Q')) \quad .$$

If x,y are two generators as above, x^τ, y^τ will be two other generators with $v'(x^\tau) = -2$ and $v'(y^\tau) = -3$, and they will satisfy $(y^\tau)^2 = (x^\tau)^3 + ax^\tau + b$ because $a^\tau = a$ and $b^\tau = b$ (τ is trivial on k). We can then compute $j_{v'}$ by means of these two generators and get $j_{v'} = j_v$.

(2.31) Definition. A function field of one variable of genus one (over k) is called an elliptic field (over k).

We have just seen that an elliptic field has a well defined invariant

$j \in k$ if the characteristic p is neither 2 nor 3. But the reader will check easily that more generally, if $p \neq 2$, one could use Legendre's equation and define j by (I.4.3). The invariant j is called the absolute invariant of the elliptic field.

(2.32) Theorem 6. The mapping $L \mapsto j = j(L) \in k$ which associates to every elliptic field L (over the algebraically closed field k of characteristic $p \neq 2$) its absolute invariant, defines a bijection between the set of k-isomorphism classes of elliptic fields and k.

Proof. We give a proof for the case $p \neq 2,3$ (and leave the case $p = 3$ as exercise). First, if two elliptic fields L and L' have the same invariant $j = j'$, they will be the function fields of the non-singular cubics of respective (affine) equations

$$y^2 = x^3 + ax + b \quad \text{and} \quad y'^2 = x'^3 + a'x' + b' \quad .$$

If both a and a' vanish, the fields are certainly isomorphic because then $b \neq 0$, $b' \neq 0$ and we can choose a sixth root t of b'/b so that the mapping $x \mapsto x' = t^2 x$, $y \mapsto y' = t^3 y$ defines an isomorphism from V to V' and from L to L'. If $a \neq 0$, take any square root t of a'/a in k so that $a'^3 = t^6 a^3$. Then $j = j'$ implies $b'^2 = t^6 b^2$ and $b' = \pm t^3 b$. Replacing t by -t if necessary, we may assume $b' = t^3 b$ and define as before an isomorphism $L \xrightarrow{\sim} L'$. Secondly, we have to see that any element j (in k) is the invariant of an elliptic field L/k. If $j = 0$ take for example for L the function field of the non-singular cubic $y^2 = x^3 + 1$. If $j \neq 0$, replace the preceeding curve by $y^2 = x^3 + x + b$ where b is any root of $\frac{j}{1728} = 4/(4 + 27b^2)$ (there are many other possible choices!).

(2.33) Remark 1. A law $\underset{Q}{+}$ with Q as neutral element can be introduced on a non-singular cubic V as before (2.7) without the assumption that Q is a flex (but $p \neq 2$). Because $Q' \mapsto Q' \underset{P}{+} Q$ defines a birational isomorphism $V \xrightarrow{\sim} V$ (i.e. an isomorphism $L \xrightarrow{\sim} L$), we see that this translation gives an isomorphism between the laws $\underset{P}{+}$ and $\underset{Q}{+}$. In particular, this last law is associative (a group law).

(2.34) Definitions for arbitrary field k. When k is not algebraically closed, we say that L is a function field of one variable over k if

a) k is algebraically closed in L ,

b) L is separable over k and of finite type ,

c) L is of transcendental degree one over k .

Then $\bar{k}$ (an algebraic closure of k) is linearly disjoint from L over k. We say that L is an elliptic function field over k when moreover

d) $L\bar{k}$ is of genus one over $\bar{k}$.

An elliptic function field has an invariant (we have defined it only when $p \neq 2$) $j \in \bar{k}$ which will be invariant under all automorphisms of $\bar{k}/k$ hence purely inseparable over k : $j \in k^{p^{-\infty}}$.

An algebraic curve E is called an elliptic curve over k if it has a point rational over k and if its function field k(E) is an elliptic field over k. Such a curve is always k-isomorphic to a plane curve in Weierstrass form : if $P \in E(k)$, take $\xi \in L(2(P)) - L((P))$, $\eta \in L(3(P)) - L(2(P))$, then $\varphi = (\xi, \eta) : E \rightarrow \mathbf{P}^2$ is rational over k, $\varphi(E)$ has a Weierstrass equation and $\varphi(P)$ (point at infinity on $\varphi(E)$ by construction) is a flex on $\varphi(E)$. If $E \subset \mathbf{P}^2$ is itself a plane curve and $P \in E(k)$ is a flex, then there is a coordinate system of $\mathbf{P}^2$ in which E has a Weierstrass equation.

(2.35) An example. Let us show that a field of genus one does not satisfy necessarily the conditions of definition (2.13) if $k \neq \bar{k}$. Let k be a non-perfect field of characteristic 3 and $a \in k - k^3$, so that $\alpha = a^{1/3} \in k^{1/3}$ is not in k. Let L be the function field k(V) where V is the cubic of (affine) equation $y^2 = x^3 - a$, so that $L = k(x, (x^3 - a)^{\frac{1}{2}})$ is the separable quadratic extension of k(x) obtained by adjoining a square root of $x^3 - a$. The cubic $y^2 = x^3 - a$ has a singular point $(\alpha,0)$ over $\bar{k}$, so that its group law (having for neutral element the flex at infinity) is defined at all points of $V_k = V(k)$ (points having coordinates in k in one representation). This enables to show that for all divisors $\underline{d} \in \mathrm{Div}(V_k)$ of degree $d > 0$, then $\dim_k L(\underline{d}) = d$ exactly as in the proof of Th.5 (use (2.6)). However, if k' is the extension $k(\alpha)$ (or $k^{1/3}$, or $\bar{k}$) of k,

$$L' = L\cdot k' = k'(x,(x-\alpha)^{3/2}) = k'((x-\alpha)^{\frac{1}{2}})$$

is a rational field, hence of genus 0 over k'. This example shows that some care must be taken with non-perfect base fields k and this is why we define the genus after the constant field extension $\bar{k}$ instead of using (2.13) directly over k. We also require E(k) not empty for an elliptic curve E over k in order to have a group structure over E(k) (and not only over $E = E(\bar{k})$).

(2.36) Remark 2. Let k be a not necessarily algebraically closed field of characteristic $p \neq 2,3$ and let

$$V : y^2 = x^3 + ax + b \quad ,$$

$$V' : y^2 = x^3 + a'x + b' \quad ,$$

be two absolutely irreducible cubics defined over k ($a,a',b,b' \in k$) and having the same absolute invariant $j = j'$. Then these curves are isomorphic over $\bar{k}$. If neither a nor b vanish, the equality $j = j'$ implies $(a'/a)^3 = (b'/b)^2 \in k$ hence the existence of a $t \in k$ with $a'/a = t^2$, $b'/b = t^3$ and the isomorphism $V \xrightarrow{\sim} V'$ given by $x \longmapsto tx$, $y \longmapsto t^{3/2}y$ is already defined over $k(t^{\frac{1}{2}})$. When $a = 0$ (case $j = j' = 0$) we have $a' = 0$ and hence $bb' \neq 0$. Take for μ a sixth root of b'/b and define the isomorphism $V \xrightarrow{\sim} V'$ by $x \longmapsto \mu^2 x$, $y \longmapsto \mu^3 y$. It is rational over the extension $k(\mu)$ of degree six of k. Similarly if $b = b' = 0$ (case $j = j' = 1728$), there is an isomorphism $V \xrightarrow{\sim} V'$ rational over an extension of degree four of k.

(2.37) Complement. We have used the result asserting that a function field of one variable over an algebraically closed field k has a separating base (consisting of one element x), to show that such a field is always isomorphic to the rational function field over an irreducible plane curve (cf. before (2.13)). This is well-known (e.g. Bourbaki, Algèbre, Chap.V, §9 n°3). We indicate here a simple proof avoiding derivations. More precisely, we prove :

Let k be a perfect field, L a finitely generated extension of k of transcendental degree one. Then there exists an element $x \in L$ such that $L/k(x)$ is algebraic separable. Let us suppose first that $L = k(x,y)$ is generated by two elements, and let $P(x,y) = 0$ with $P \in k[X,Y]$ be an algebraic dependence relation between x and y. Let f and ℓ be the maximal positive integers such that

$$P(X,Y) = Q(X^{p^f}, Y^{p^\ell}) .$$

We show that $L/k(x)$ is algebraic separable if $\ell \leqslant f$ (and $L/k(y)$ algebraic separable if $f \leqslant \ell$). Because $k = k^p$, we can write

$$P(X,Y) = Q'(X^{p^{f-\ell}}, Y)^{p^\ell} ,$$

and since $P(x,y) = 0$, we shall have $Q'(x^{p^{f-\ell}}, y) = 0$. This equation is separable for y with coefficients in $k(x^{p^{f-\ell}}) \subset k(x)$. Then the general case is treated by induction. If $L = k(y_1,\ldots y_{n+1})$, assume that $y_1,\ldots,y_n$ are separable over $k(x)$. Let P as above express an algebraic relation between x and y_{n+1}. By extraction of pth roots, it will furnish a separable relation for x over $k(y_{n+1})$ or for y_{n+1} over $k(x)$. The result follows by the transitivity property of separability.

The reader will note that this property fails in general if we drop the assumption of finite generation : take k algebraically closed of characteristic $p \neq 0$ and an indeterminate X; then $L = k(X^{p^{-f}})_{f \geqslant 0}$ has no element x with $L/k(x)$ algebraic and separable.

3. Differential forms and elliptic integrals

In this section, we work throughout with an algebraically closed field k of characteristic p = 0 (in the second part, where we consider elliptic integrals and differential equations, we shall even assume $k = \mathbb{C}$).

We start by giving an algebraic definition of differentials on curves (globally, without speaking of the line bundles over the curve over which they can be interpreted as sections).

(3.1) Definition. Let K be a function field of one variable over k and E a K-vector space. Then we denote by $Der_k(K,E)$ the K-vector space of derivations $D : K \longrightarrow E$ with values in E and trivial on k. Such derivations are mappings satisfying

a) $D(a) = 0$ for all $a \in k$,

b) $D(f_1 + f_2) = D(f_1) + D(f_2)$ for all $f_1, f_2 \in K$,

c) $D(f_1 f_2) = f_1 D(f_2) + f_2 D(f_1)$ for all $f_1, f_2 \in K$.

In particular they are linear mappings over k because a) and c) imply that $D(af) = aD(f)$ for $a \in k$ and $f \in K$.

(3.2) Example. If $K = k(x)$ is a rational field, the mapping $D \longmapsto D(x)$ is a bijection $Der_k(K,E) \longrightarrow E$. This follows simply from the fact that the formal rules of derivations for quotients are true and so D(f) is determined, as soon as D(x) is given arbitrarily in E , for all rational expressions $f = f(x)$ in x.

(3.3) Proposition. If K' is a finite algebraic extension of the function field of one variable K over k, then the restriction operation $Der_k(K',E) \longrightarrow Der_k(K,E)$ is bijective. (In other words, a k-derivation D of K has a unique extension as k-derivation of K'.)

<u>Proof</u>. By the primitive element theorem $K' = K(y)$ for an element $y \in K'$ (we are in characteristic 0). Let $P(y) = 0$ be the minimal polynomial of y over K. For all extensions $\bar{D}$ of D we must have $P^D(y) + P'(y)\cdot\bar{D}(y) = 0$ where P^D is the polynomial obtained from P by replacing its coefficients by their derivatives, and P' is the usual derivative of P. Since y is simple root of P (separability), we have $P'(y) \neq 0$ and so $\bar{D}(y)$ is necessarily given by $-P^D(y)/P'(y)$. This proves the uniqueness of the extension $\bar{D}$. The existence follows from the definition $\bar{D}(y) = -P^D(y)/P'(y)$. This certainly defines a derivation of the polynomials ring $K[Y]$ which will be trivial on the principal ideal generated by P, which is the ideal of algebraic relations satisfied by y (incredulous readers may have to check it pen in hand, or go back to Bourkaki, Algèbre, Chap. V, §9 n°1 Prop.5), and hence a derivation of $K' = K[Y]/(P(Y))$.

(3.4) <u>Corollary</u> 1. <u>For every function field</u> K (<u>of one variable over</u> k) <u>and any</u> K-<u>vector space</u> E, $Der_k(K,E)$ <u>is isomorphic to</u> E <u>over</u> K.

<u>Proof</u>. Let x be any element of K - k. Then K is a finite algebraic extension of the rational field $k(x)$. An isomorphism is then given by

$$Der(K,E) \longrightarrow Der(k(x),E) \longrightarrow E$$
$$D \longmapsto \text{restriction of } D \longmapsto D(x) \ .$$

Note that the composite of these two bijections is indeed K-linear (in $Der(k(x),E)$ we consider E as $k(x)$-vector space by restriction of the scalars). Also note that there is <u>no canonical</u> isomorphism between the spaces in question.

(3.5) <u>Corollary 2</u>. <u>If</u> $f,g \in K$ <u>and</u> $g \notin k$, <u>then</u> $D(f)/D(g) \in K$ <u>is independent of the non-zero derivation</u> $D \in Der(K) = Der_k(K,K)$.

<u>Proof</u>. Observe that if $g \notin k$, $D(g) = 0$ only for $D = 0$ (this is the only extension of the zero derivation of $k(g)$ to K). Because $Der(K)$

is of dimension one over K, the quotients have a well-defined meaning. Then, if D' is another non-zero derivation, D' = hD for some $h \in K$ proves that $D'(f)/D'(g) = D(f)/D(g)$.

As in the proof of this corollary we shall denote $Der_k(K,K)$ simply by $Der_k(K)$ or by Der(K) if there is no risk of confusion about which field k we are using. Also, if $g \in K - k$ we denote by $D_g \in Der(K)$ the unique derivation defined by $D_g(g) = 1$. The above corollary can be reformulated as $D(f)/D(g) = D_g(f) \in K$ is independent of the derivation $D \neq 0$ chosen in Der(K).

(3.6) Definition. Let K and E be as in (3.1). We define the space of E-valued differentials of K, in notation $Diff_k(K,E)$, to be the K-dual of the space $Der_k(K,E)$. Thus, an E-valued differential form of K is a linear form $\omega : Der_k(K,E) \longrightarrow K$.

By Cor.1 above, $Diff_k(K) = Diff_k(K,K)$ is a vector space of dimension one over K (with no canonical basis, so that we cannot multiply its elements, but we can divide any element by a non-zero one, finding a quotient in K).For example, each $f \in K$ defines a differential form $df : D \longmapsto \langle D,df \rangle = D(f)$. This gives a k-linear mapping

(3.7) $\quad d : K \longrightarrow Diff_k(K) , \ f \longmapsto df$.

One can check easily that this mapping is in fact a derivation :

$$\langle D,d(fg)\rangle = D(fg) = fDg + gDf = f\langle D,dg\rangle + g\langle D,df\rangle =$$
$$= \langle D,fdg + gdf\rangle \text{ for every } D \in Der(K) .$$

One has df = 0 exactly when D(f) = 0 for all derivations D, hence precisely when $f \in k$. This shows that $Ker(d) = k \subset K$. Although K and $Diff_k(K)$ are one dimensional K-vector spaces, d is not surjective (as we shall see) because it is only k-linear (and not K-linear). A differential of the form df (for some $f \in K$) is said to be exact, so that d(K) is the k-subspace of exact differentials in $Diff_k(K)$.

There is no canonical derivation D in $\mathrm{Der}_k(K)$, but d is a canonical derivation of K with values in $\mathrm{Diff}_k(K)$. More precisely :

(3.8) *Proposition*. *The canonical derivation* $d : K \longrightarrow \mathrm{Diff}_k(K)$ *is universal in the sense that every E-valued derivation* $D : K \longrightarrow E$ *can be factored uniquely through* d , *with a K-linear mapping* $A_D : \mathrm{Diff}_k(K) \longrightarrow E$, *i.e. so that the following diagram is commutative*

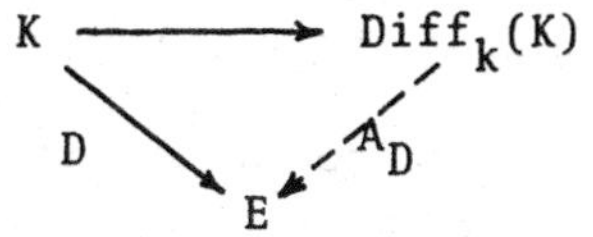

Proof. The *uniqueness* of the factorization is clear because $d(K) \neq 0$ generates the K-vector space $\mathrm{Diff}_k(K)$ of dimension one. For the *existence*, take any $x \in K - k$. There exists a unique K-linear map $A_D : \mathrm{Diff}_k(K) \longrightarrow E$ such that $A_D(dx) = D(x)$ (note that $dx \neq 0$ generates $\mathrm{Diff}_k(K)$ over K, and that A_D depends - a priori - on x). Then it is clear that $A_D \circ d$ is a derivation (over k). The derivation $D - A_D \circ d \in \mathrm{Der}_k(K,E)$ is zero on x by definition, hence is identically zero (unique extension of the zero derivation $k(x) \longrightarrow E$ (3.3)). This proves that A_D gives a factorization of D (and is independent of the special choice of x by the uniqueness already proved).

Let f and $g \in K$ with g not constant ($g \notin k$). Then df and dg must be proportional in $\mathrm{Diff}_k(K)$ with a coefficient in K. This coefficient may be computed by comparing one (any) value of these linear forms

$$df/dg = \langle D,df \rangle / \langle D,dg \rangle = D(f)/D(g) = D_g(f) \quad (0 \neq D \in \mathrm{Der}(K)) .$$

In particular if we fix a transcendental element $x \in K$ (over k) ,

(3.9) $\qquad df/dx = D_x(f)$ for every $f \in K$,

so that the differential quotient df/dx coincides with the derivation with respect to x : $d/dx = D_x$ is the formal derivative with respect to x.

By choosing generators x,y of K, we can assume that $K = k(x,y)$ is the field k(V) of k-rational functions over a plane curve V defined over k.

(3.10) Lemma. Let $P \in V$ be a regular point and $\pi = \pi_P$ a generator of the maximal ideal M_P of the local ring R_P at P : $\mathrm{ord}_P(\pi) = 1$. Then

$$df/d\pi = D_\pi(f) \in R_P \quad \text{for all } f \in R_P .$$

Proof. The expansion of functions $f \in K$ in Laurent series at P gives an embedding $K \longrightarrow k((\pi))$ of the field $K = k(V)$ into the field of all formal Laurent series in π with coefficients in k (this is the field of fractions of the ring $k[[\pi]]$ of integral formal series in π and coefficients in k). The derivation term by term with respect to π of formal Laurent series gives a derivation $\widetilde{D}_\pi \in \mathrm{Der}_k(k((\pi)))$ which coïncides with $D \in \mathrm{Der}_k(K)$ on the element $\pi \in K$. By (3.3), $\widetilde{D}_\pi$ must coïncide with D_π on the algebraic subextension K of $k(\pi)$ in $k((\pi))$. Since elements $f \in R_P$ are characterized by the fact that they have an image in $k[[\pi]]$, their derivative $D_\pi(f) = \widetilde{D}_\pi(f)$ will also be in $k[[\pi]]$. Also, one sees from $D_\pi = \widetilde{D}_\pi$ that if π' is another uniformizing variable at P, that $d\pi'/d\pi = \widetilde{D}_\pi(\pi') \in R_P^\times$ is a unit of the local ring R_P : $(d\pi'/d\pi)(P) \in k^\times$.

This lemma shows that $d/d\pi = D_\pi$ is necessarily continuous on R_P for the topology defined by taking the powers M_P^r of the maximal ideal M_P as basis of neighbourhoods of 0 in R_P. Consequently, Laurent series may be derived term by term.

(3.11) Definition. Let P be a regular point of V, and π such that $\mathrm{ord}_P(\pi) = 1$. If $\omega = f d\pi \in \mathrm{Diff}_k(K)$ is a differential, we define its order at P by $\mathrm{ord}_P(\omega) = \mathrm{ord}_P(f)$ (when defined : $f \neq 0$ or $\omega \neq 0$).

This is a meaningful definition because if π' is another uniformizing variable at P, $d\pi/d\pi'$ is a unit in R_P and thus $\omega = f d\pi = f(d\pi/d\pi')d\pi'$, and $\mathrm{ord}_P(f) = \mathrm{ord}_P(f \cdot d\pi/d\pi')$.

Properties which follow immediately from the definition are

(3.12) a) $\mathrm{ord}_P(\omega + \omega') \geqslant \mathrm{Inf}(\mathrm{ord}_P(\omega), \mathrm{ord}_P(\omega'))$

(with equality if these orders are distinct) ,

b) $\mathrm{ord}_P(f\omega) = \mathrm{ord}_P(f) + \mathrm{ord}_P(\omega)$ for $f \in K$, $\omega \in \mathrm{Diff}(K)$.

From the proof of (3.10) we can also see that

(3.13) c) $\mathrm{ord}_P(df) = \mathrm{ord}_P(f) - 1$ if $\mathrm{ord}_P(f) \neq 0$ $(f \in K)$.

The points $P \in V$ where a differential ω has order <0 are called poles of ω and the points P where it has order >0 are called zeros of ω.

(3.14) Proposition. *Let K be the function field of k-rational functions over a non-singular plane curve V defined over k, and $\omega \neq 0$ be a differential form of K. Then* $\mathrm{div}(\omega) = \sum_{P \in V} \mathrm{ord}_P(\omega)(P) \in \mathrm{Div}(V)$. *In other words, a non-zero differential of K has only finitely many zeros and poles.*

Proof. By choice, x is transcendant over k, so that V is not a vertical line in the (X,Y) coordinate system and $dx = d(x - x(P))$ is the differential of a uniformizing variable at all P with non-vertical tangent (this last set is finite by (1.10)). Thus if $\omega = f dx$, we shall have $\mathrm{ord}_P(\omega) = \mathrm{ord}_P(f)$ at nearly all points P and this is zero except if P appears in the divisor of f.

(3.15) Definition. *A differential form $\omega \in \mathrm{Diff}_k(K)$ is said to be abelian, or of first kind when it has no pole : $\mathrm{ord}_P(\omega) \geqslant 0$ for all points $P \in V$ (which is supposed non-singular).*

The abelian differentials make up a vector space $\mathfrak{D}_1$ over k. An exact differential df can be abelian only when $f \in k$ and $df = 0$ (3.13).

An example of non-zero abelian differential is given by the following

(3.16) Proposition. *The dimension (over k) of the space $\mathfrak{D}_1$ of abelian differentials of K is 0 if K is of genus 0 over k (K rational over k) and 1 if K is of genus 1 over k.*

Proof. Take first the case $K = k(x)$ is rational over k. Then V is the projective line $k \cup \{\infty\}$ (identified with the X-axis). Because $x - x(P)$ is a uniformizing variable at all finite points P and $dx = d(x - x(P))$, we see that the condition $\mathrm{ord}_P(fdx) \geqslant 0$ for all finite points P implies that $f \in k[x]$ is a polynomial in x. At the point at infinity, we have $\mathrm{ord}_\infty(x) = -1$ so that we can take the uniformizing variable $\pi_\infty = 1/x$. Then $x = 1/\pi_\infty$, $dx = -\pi_\infty^{-2} d\pi_\infty$ and $\omega = fdx = -f(1/\pi_\infty)/\pi_\infty^2 d\pi_\infty$ has order $\mathrm{ord}_\infty(\omega) \leqslant -2$ if $f \neq 0$. This shows $\mathfrak{D}_1 = 0$. We turn to the case of genus one and suppose $K = k(x,y)$ is the function field over the cubic of equation $y^2 = x^3 + ax + b$ (for some $a, b \in k$). I claim that $\omega = dx/y = (1/y)dx$ has order 0 at all points P of the cubic V. This is obvious if P is not among the E_i ($i = 0,\dots,3$, points of order 1 or 2) because then $x - x(P)$ is a uniformizing variable at P and $\mathrm{ord}_P(dx/y) = \mathrm{ord}_P(d(x - x(P))) - \mathrm{ord}_P(y) = \mathrm{ord}_P(x - x(P)) - 1 - 0 = 0$ (because $y(P) \neq 0$ at these points). At the points $E_i = (e_i, 0)$ $(i=1,2,3)$ we have $\mathrm{ord}_i(y) = 1$ and $\mathrm{ord}_i(dx) = \mathrm{ord}_i(d(x - e_i)) = 2 - 1 = 1$ which gives $\mathrm{ord}_i(dx/y) = 0$. At the flex at infinity E_0 (neutral element), we choose the uniformizing parameter $\pi_\infty = y/x^2$ (2.21) (we could also choose x/y). Small computations give

$$d\pi_\infty = (1/x^2)dy - 2(y/x^3)dx = (2ydy)/2x^2y - 2(x^3 + ax + b)/x^3 \cdot dx/y =$$
$$= (-1/2 - 3a/(2x^2) - 2b/x^3) \cdot dx/y .$$

We can write this in the form

(3.17) $\qquad d(y/x^2) = f \cdot dx/y$ with $f \in R_\infty^\times$ and $f(\infty) = -\frac{1}{2}$.

This proves in particular $\mathrm{ord}_\infty(dx/y) = 0$. Had we only needed this order, we could have proceeded more expediently as follows (using (2.21) and (3.13))

$$\mathrm{ord}_\infty(dx/y) = \mathrm{ord}_\infty(dx) - \mathrm{ord}_\infty(y) = \mathrm{ord}_\infty(x) - 1 + 3 = 0 ,$$

but we shall use the more precise form given in (3.17). The conclusion of the proof is now straightforward : if $\omega' = f(dx/y)$ is an abelian differential form, we must have $\mathrm{div}(f) = 0$, so that $f \in k$.

We are now able to give a purely algebraic definition of the residue of a differential form at a regular point P.

(3 18) Proposition. Let K be the function field over the plane curve V, and let P be a regular point on V. Then there exists a unique linear form

$$\mathrm{res}_P : \mathrm{Diff}_k(K) \longrightarrow k \quad ,$$

having the following properties :

a) $\mathrm{res}_P(\omega) = 0$ if $\omega \in \mathrm{Diff}_k(K)$ with $\mathrm{ord}_P(\omega) \geqslant 0$,

b) $\mathrm{res}_P(df) = 0$ for all exact differentials $(f \in K)$,

c) $\mathrm{res}_P(\frac{df}{f}) = \mathrm{ord}_P(f)$ for all $f \in K$.

Proof. Let $\pi = \pi_P$ be a uniformizing variable at P : $\mathrm{ord}_P(\pi) = 1$, and write the Laurent series in π of a function $f \in K$ as follows :

$$f = \sum_{i<-1} a_i \pi^i + a_{-1}/\pi + f_o \quad \text{with } f_o \in R_P \quad .$$

The differential $\omega = f d\pi$ can then be written

$$\omega = \sum_{i<-1} a_i \pi^i d\pi + a_{-1} d\pi/\pi + f_o d\pi \quad .$$

The conditions b),a),c) in this order give then respectively

$$\mathrm{res}_P(\pi^i d\pi) = \mathrm{res}_P\, d(\pi^{i+1}/(i+1)) = 0 \quad (i < -1) \ ,$$

$$\mathrm{res}_P(f_o d\pi) = 0 \quad ,$$

$$\mathrm{res}_P(a_{-1} d\pi/\pi) = a_{-1}\mathrm{ord}_P(\pi) = a_{-1} \quad (k\text{-linearity}) \quad .$$

Any k-linear form having the three required properties will thus give $\mathrm{res}_P(f d\pi) = a_{-1}$, and this shows the uniqueness. Conversely, the existence is proved by defining $\mathrm{res}_P(f d\pi) = a_{-1}$ (with a choice of uniformizing variable π) and by checking that the three properties are indeed satisfied. For the last one, write $f = \pi^n g$ with $g \in R_P^{\times}$ (hence $n = \mathrm{ord}_P(f)$) and take the logarithmic derivative ! This definition is independent of the special uniformizing parameter selected by the previous proof of uniqueness.

By definition, the residue of a differential form ω of K at the regular point $P \in V$ is the constant $\mathrm{res}_P(\omega) \in k$.

(3.19) Definition. Let V be a non-singular (irreducible, plane) curve and K its function field k(V). A differential form $\omega \in \mathrm{Diff}_k(K)$ is said to be of second kind when $\mathrm{res}_P(\omega) = 0$ for all points $P \in V$, and of third kind when $\mathrm{ord}_P(\omega) \geq -1$ for all points $P \in V$.

We shall denote by $\mathfrak{D}_2$ the k-subspace of $\mathrm{Diff}_k(K)$ of forms of second kind and by $\mathfrak{D}_3$ the k-subspace of $\mathrm{Diff}_k(K)$ of forms of the third kind. These classes are by no means mutually disjoint. For example a differential of first kind is also of second and third kind. Also note that by definition, exact differentials are of second kind. The following formulas hold :

(3.20) $d(K) \subset \mathfrak{D}_2$, $d(K) \cap \mathfrak{D}_1 = \{0\}$, $\mathfrak{D}_1 = \mathfrak{D}_2 \cap \mathfrak{D}_3$.

(3.20) Examples. 1) We consider a non-singular cubic of (affine) equation $y^2 = x^3 + ax + b$ $(4a^3 + 27b^2 \neq 0)$ and we take the differential $\omega = xdx/y$. Only the point E_0 at infinity deserves special consideration. We take there the uniformizing variable $\pi = \pi_\infty = y/x^2$, so that $x\pi^2 = y^2/x^3 = 1 + a/x^2 + b/x^3 = 1 + \pi^4 u \in 1 + M_\infty^4$. Also $x \in \pi^{-2} + M_\infty^2$ has no term in π^{-1} . Using now (3.17)

$$xdx/y = (-2\pi^{-2} + v)d\pi \quad \text{with some regular } v \in R_\infty ,$$

which shows that xdx/y has no residue at ∞ hence no residue at all. This is consequently a differential of second kind on the elliptic curve V.

2) The differential form $dx/(x - x_0)$ (with any $x_0 \in k$) on the same elliptic curve as before is a third kind differential. Indeed,

$$\mathrm{ord}_\infty(dx/(x - x_0)) = \mathrm{ord}_\infty(y/(x - x_0)) + \mathrm{ord}_\infty(dx/y) =$$
$$= -3 + 2 + 0 = -1 \quad .$$

At a point $P \in V$ where $x(P) = x_0$, $x - x_0$ has a simple zero if $P \neq -P$ and a double zero if $P = -P$ in which case y has a simple zero. This shows that the differential in question has at most simple poles and is therefore of third kind. More precisely we have seen that its divisor is

$$\mathrm{div}(dx/(x - x(P))) = -(\infty) - (P) - (-P) + (E_1) + (E_2) + (E_3) .$$

The interested reader can check that its residues are given by

$$\mathrm{res}_P(dx/(x - x(P))) = \mathrm{res}_{-P}(\ldots) = 1 , \ \mathrm{res}_\infty(\ldots) = -2 .$$

Another differential of the third kind is given by $(x - x(P))^{-1}dx/y$ when $P \neq E_i$. Its divisor is indeed equal to the divisor of $(x - x(P))$ which has only simple poles when P is not on the X-axis. Its residues are given by $\mathrm{res}_P(\omega) = 1/y(P)$ and $\mathrm{res}_{-P}(\omega) = -1/y(P)$.

Now we restrict ourselves to the consideration of non-singular elliptic curves.

(3.21) <u>Lemma</u>. <u>Let</u> V <u>be a non-singular cubic of equation</u> $y^2 = x^3 + ax + b$ <u>and let</u> K <u>be its function field</u> k(V) $(a,b \in k$ <u>and</u> $4a^3 + 27b^2 \neq 0)$. <u>Then</u> $\sum_{P\in V} \mathrm{res}_P(\omega) = 0$ <u>for all differential forms of</u> K .

<u>Proof</u>. Every $f \in K = k(x,y)$ can be written uniquely in the form $f = a(x) + b(x)y$ with rational functions $a(x), b(x) \in k(x)$. We observe then that for any point $P \in V$, $\mathrm{res}_P(a(x)dx) = \mathrm{res}_{-P}(a(x)dx)$ and $\mathrm{res}_P(b(x)ydx) = -\mathrm{res}_{-P}(b(x)ydx)$, so that grouping these terms together $\mathrm{res}_P(fdx) + \mathrm{res}_{-P}(fdx) = \mathrm{res}_P(f + f^\sigma)dx = 2\mathrm{res}_P(a(x)dx)$ where we have written $f^\sigma = a(x) - b(x)y$ (σ is the non-trivial automorphism of the quadratic extension K of k(x), corresponding to the symmetry $P \longmapsto -P$ of V). Then we can write

$$\sum_{P\in V} \mathrm{res}_P(fdx) + \sum_{-P\in V} \mathrm{res}_{-P}(fdx) = 2\sum_{P\in V} \mathrm{res}_P(a(x)dx) .$$

But $\mathrm{res}_P(a(x)dx) = \mathrm{res}_{x(P)}(a(x)dx)$ if we consider now a(x) as function on the line Y = 0 (X-axis). We are reduced to proving the lemma for a rational field. There we use the canonical decomposition of a(x) in principal parts at its poles $\neq \infty$ and a polynomial (this is an additive decomposition). As the lemma is satisfied for all these principal parts (only two opposite residues occur, with one of them at infinity) and for a polynomial (no residue at all), it is proved.

This proof of the lemma can in fact be generalized for any function field (over a non-singular plane curve with our definitions). Because $\operatorname{div}(dx/y) = \underline{0}$, we see that $\operatorname{div}(f dx/y) = \operatorname{div}(f)$ and so, all divisors of differential forms are principal divisors. They are in particular of degree 0. This is a special feature of elliptic curves. One can check easily that the degree of the divisor of any differential form of the rational field $k(x)$ is -2. In general this degree is $2g-2$ on a curve of genus g : it vanishes precisely for elliptic curves. We give now Abel's classification of differential forms on elliptic curves.

(3.22) Theorem (Abel). Let K be a function field of one variable over k (algebraically closed and of characteristic 0) of genus one, and choose generators $x, y \in K$ so that $K = k(x,y)$ with $y^2 = x^3 + ax + b$. Then every differential form $\omega \in \operatorname{Diff}_k(K)$ can be written uniquely in the form

$$\omega = \underbrace{df + \alpha x dx/y + \beta dx/y}_{\mathfrak{D}_2} + \sum_{P \in V_0} \gamma_P \frac{y + y(P)}{x - x(P)} \frac{dx}{y}$$

with $\mathfrak{D}_1$ = the term $\beta dx/y$ and $\mathfrak{D}_3$ = the terms $\beta dx/y + \sum_{P \in V_0} \gamma_P \frac{y + y(P)}{x - x(P)} \frac{dx}{y}$,

with $f \in K$, $\alpha, \beta, \gamma_P \in k$.

(The components in the $\mathfrak{D}_i$'s are not uniquely determined, but this gives a possible decomposition.)

Proof. Consider the differential

$$\omega_2 = \omega - \sum_{P \in V_0} \gamma_P \frac{y + y(P)}{x - x(P)} \frac{dx}{y} \quad \text{with} \quad \gamma_P = \tfrac{1}{2} \operatorname{res}_P(\omega)$$

(with a summation on the affine part $V_0 = V - \{E_0\}$ of V). I claim that ω_2 is of second kind. For this we observe that the zeros of $x - x(P)$ occur at P and $-P$ with corresponding $y(-P) = -y(P)$: $(y + y(P))/(x - x(P))$ has a simple pole at P and no pole at $-P$ (if $-P \neq P$). The function $y^{-1}(y + y(P))$ takes the value 2 at P and the value 0 at $-P$ (if $-P \neq P$) with multiplicity one. This shows that the residue at P is 2 (no pole at -P if distinct from P). By (3.21) we see that the residue at infinity must be -2 (one could also compute it directly using (3.17)). Now by definition, ω_2 has no residue at all finite points $P \in V_0$ and consequently no residue at infinity by lemma (3.21). We are reduced to proving the theorem for second kind differentials. Then we write $\omega = \omega_2 = f dx/y$ with $f = a(x) + b(x)y$, and by additive decomposition of a(x) and b(x) according to the principal parts of their poles (with polynomials corresponding to the poles at infinity), we get a linear combination of differentials

$$x^n dx/y \ , \ x^n dx = d(\frac{x^{n+1}}{n+1}) \ , \ (x - \xi)^{-n} dx/y \ , \ (x - \xi)^{-n} dx$$

with positive integers n. The last term is an exact differential except for n = 1 when it is a third kind differential. The term before, $(x - \xi)^{-n} dx/y$, is of second kind for $n > 1$ (because x is an even function...) and for n = 1 it is of third kind if $\xi \neq e_i$ (i=1,2,3) and of second kind otherwise. If we start with a second kind differential, we may disregard the terms which are not of second kind in the decomposition, because they will eventually compensate each other. Let us consider the more typical term $x^n dx/y$ (which is of second kind with a pole of order 2n at infinity E_0). To reduce it to the case n = 1, we make the following computation :

$$\begin{aligned} d(x^n y) &= n x^{n-1} y dx + x^n dy = \\ &= n x^{n-1}(x^3 + ax + b) dx/y + \tfrac{1}{2} x^n (3x^2 + a) dx/y = \\ &= \left[(n + \tfrac{3}{2}) x^{n+2} + a(n + \tfrac{1}{2}) x^n + b x^{n-1} \right] dx/y \ . \end{aligned}$$

This formula shows how one can compute $x^{n+2} dx/y$ in terms of $x^n dx/y$, $x^{n-1} dx/y$, $d(x^n y)$. If n = 0 , the coefficient of $x^{n-1} dx/y$ is 0 so that this term disappears. By descending induction, all these terms are brought down to combinations of exact differentials, and of dx/y , $x dx/y$. This gives the full treatment for the pole at infinity of ω_2 . Other poles can be treated similarly or brought at infinity after a change of variables (a translation of the curve). The uniqueness of this special decomposition comes from the fact that the γ_P are uniquely determined (they are the half residues of ω) and then an equality $df = \alpha xdx/y + \beta dx/y$ would imply that f has only one pole, a simple one at infinity, which is impossible in a field of genus $\neq 0$, unless $df = 0$, $\alpha = 0$, $\beta = 0$.

(3.23) <u>Corollary</u>. $\mathfrak{D}_2/(\mathfrak{D}_1 + d(K))$ <u>is of dimension one over</u> k.

<u>Proof</u>. Indeed the differential xdx/y generates a supplement from $\mathfrak{D}_1 + d(K)$ in $\mathfrak{D}_2$.

This classification theorem describes exactly the deviation from exactness for differential forms over a field of genus one. In particular, the k-codimension of $\mathfrak{D}_2$ in $\mathrm{Diff}_k(K)$ is infinite.

Let us now turn to the integration of differentials over an elliptic curve defined over $k = \mathbb{C}$. According to the preceding classification, there are three kinds of elliptic integrals.

(3.24) <u>Elliptic integral of the first kind</u>.

Let us take an elliptic curve $y^2 = 4x^3 - g_2x - g_3$ with complex coefficients g_2, g_3 satisfying $g_2^3 - 27g_3^2 \neq 0$. We consider the projective complex curve $V = V_{\mathbb{C}}$ in the projective plane $\mathbb{P}^2(\mathbb{C})$, and we use its parametrization by means of the functions $\wp$ and $\wp'$ of Weierstrass

$$\varphi = (\wp, \wp') : \mathbb{C} \longrightarrow V \subset \mathbb{P}^2(\mathbb{C})$$

(3.25)

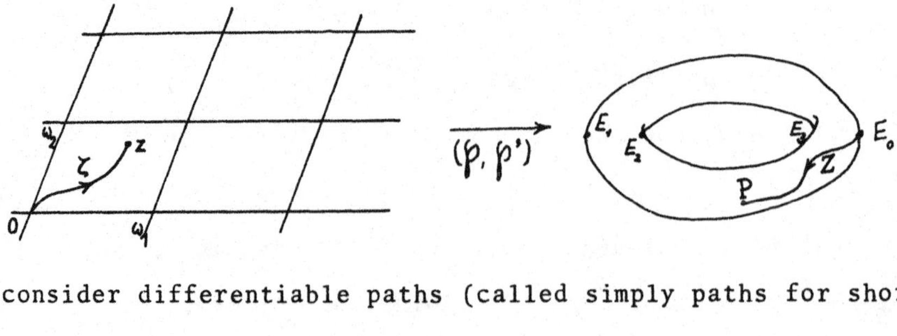

and we consider differentiable paths (called simply paths for short) on V which are images of paths in $\mathbb{C}$. Let $\zeta : [0,1] \longrightarrow \mathbb{C}$ be any path starting at the origin $\zeta(0) = 0$ with end point $z = \zeta(1)$, and let Z denote its image $Z = \varphi \circ \zeta$ as path in V. This path starts at the flex E_0 at infinity on V and ends at a certain point $P = Z(1) \in V$. Then we can compute the line integral $\int_Z dx/y$ of the first kind differential dx/y by pull back by means of φ:

$$(3.26)\quad \int_Z dx/y = \int_\zeta d\wp(u)/\wp'(u) = \int_\zeta du = \zeta(1) - \zeta(0) = z .$$

Let ω_1 , ω_2 be a direct basis of the lattice of periods of these functions of Weierstrass (corresponding to g_2 and g_3) and Ω_i be the linear path $t \longmapsto t\omega_i$ ($t \in [0,1]$). The image of Ω_i under φ is a circuit Z_i in V. The corresponding integrals

$$(3.27)\quad \int_{Z_i} dx/y = \omega_i$$

are the periods of the differential dx/y along the circuit Z_i .

Thus the integral

$$\int_{E_0}^{P} dx/y = z + n_1\omega_1 + n_2\omega_2$$

is defined up to an integral combination of the periods (depending on the chosen path Z from E_0 to P on V. The mapping

$$(3.28)\quad Z \longrightarrow \int_Z dx/y = z \quad : \quad \tilde{V} \longrightarrow \mathbb{C}$$

gives thus an isomorphism from the universal covering $\tilde{V}$ of V to $\mathbb{C}$. In this sense, the integral of the first kind on V uniformizes V.

These formulas can also be written in the form

$$(3.29)\quad \int_\infty^x (4t^3 - g_2t - g_3)^{-\frac{1}{2}}dt = (\mathrm{Arc}\,\wp)(x) \quad ,$$

or

$$(3.29)'\quad \int_\infty^{\wp(z)} (4t^3 - g_2t - g_3)^{-\frac{1}{2}}dt = z \quad ,$$

which are to be compared with

$$\int_0^x (1 - t^2)^{-\frac{1}{2}}dt = \mathrm{Arcsin}(x) \ , \quad \int_0^{\sin\psi} (1 - t^2)^{-\frac{1}{2}}dt = \psi \quad .$$

(3.30) <u>Elliptic integral of the second kind</u>.

Since the differential form xdx/y has a pole at E_0, we have to fix a starting point P_0 for our paths on V distinct from E_0. We take $P_0 = \varphi(z_0)$ with any point $z_0 \in \mathbb{C}$ not in the lattice $L = \varphi^{-1}(E_0)$. Then we take paths ζ in $\mathbb{C} - L$ starting at z_0 and ending at z. If we call $Z = \varphi \circ \zeta$ the images of these paths as before, then for any one of them

$$(3.31) \qquad \int_Z xdx/y = \int_\zeta \wp(u)\,du = \xi(z) - \xi(z_0) \ ,$$

with the primitive ξ of $\wp$ defined in (I.1.18), and this is independent of the chosen path ζ linking z_0 to z (avoiding the lattice points). The periods of this second kind differential are easily computed by taking the translated paths $\Omega_i' = z_0 + \Omega_i$ (we suppose that z_0 is so chosen that these paths avoid the lattice points) and their images in V, the circuits Z_i'. By (I.1.19) we get

(3.32)

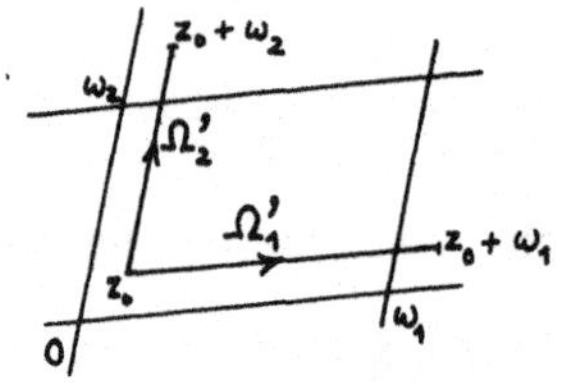

$$\int_{Z_i'} xdx/y = \xi(z_0+\omega_i) - \xi(z_0) = -\eta_i \ .$$

We can also give the less precise form to (3.31) corresponding to (3.29)

$$(3.33) \qquad \int_{\wp(z_0)}^{\wp(z)} (4t^3 - g_2t - g_3)^{-\frac{1}{2}}\,t\,dt = \xi(z) - \xi(z_0) \ ,$$

which is determined up to an integral multiple of the periods η_i (depending on the chosen path connecting P_0 to P on V).

(3.34) <u>Elliptic integral of the third kind</u>.

The third kind differential $\frac{1}{2}\,\frac{y + y(P)}{x - x(P)}\,\frac{dx}{y}$ has two simple poles, in P and E_0 with respective residues +1,-1. We make a cut S on V starting at E_0 and ending at P. We consider paths ζ in $\mathbb{C} - \bigcup S_\omega = \varphi^{-1}(V - S)$ and their images $Z = \varphi \circ \zeta$ in V - S. Then the line integral of the considered third kind differential along Z depends only on the end points $z_1 = \zeta(1)$ and $z_2 = \zeta(2)$ of ζ (because the period of the differential around the cut on V vanishes). Let P_1 and P_2 be the images of z_1 resp. z_2 in V.

(3.35) $$\int_Z \frac{1}{2} \frac{y + y(P)}{x - x(P)} \frac{dx}{y} = \int_\zeta \frac{1}{2} \frac{\wp'(u) + \wp'(z)}{\wp(u) - \wp(z)} du \quad .$$

We have to go back to (I.1.23) where we find an expression for $\wp(u) - \wp(z)$ which we derive logarithmically with respect to u and symmetrize. After replacing z by -z we find the expression

(3.36) $$\frac{1}{2} \frac{\wp'(u) + \wp'(z)}{\wp(u) - \wp(z)} = -\zeta(u+z) + \zeta(u) - \zeta(z) \quad ,$$

of which a primitive is $\log \frac{\sigma(u+z)}{\sigma(u)\sigma(z)} - \zeta(z)u$. A uniform branch of this function can be determined in $u \in \mathbb{C} - \bigcup s_\omega$ and from there, an "explicit" expression for the third kind integral can be given. However, we shall not pursue the computations any further , in particular, we do not compute the periods of this third kind integral.

(3.37) Remark. It is interesting to write the bilinear relations in integral form (cf.(I.1.20)); $\eta_1\omega_2 - \eta_2\omega_1 = 2\pi i$ gives here

$$\oint_{Z_1'} dx/y \oint_{Z_2'} xdx/y - \oint_{Z_2'} dx/y \oint_{Z_1'} xdx/y = 2\pi i \quad .$$

If we decide to associate the intersection number +1 to the two cycles Z_1' , Z_2' (in this order) on V (they have only one intersection point if they are chosen as in the figure (3.32)) we see that we can extend this definition to any couple Z,Z' of cycles on V (both are integral linear combinations of circuits) avoiding E_0 by the formula

$$Z \wedge Z' = \frac{1}{2\pi i}\left\{\oint_Z dx/y \oint_{Z'} xdx/y - \oint_{Z'} dx/y \oint_Z xdx/y\right\} .$$

Thus, in some sense, the canonical bilinear form (I.2.1) has an interpretation in terms of algebraic intersections: $B(\omega, \omega')$ is the intersection number of the two cycles on V corresponding to ω and ω' (or nearby cycles avoiding E_0).

(3.38) <u>Elliptic differential equation</u>. There is a way of defining elliptic functions by means of differential equations, which is essentially equivalent to the consideration of elliptic integrals of the first kind. The computations below will also show how an integral of the form $\int P(x)^{-\frac{1}{2}}dx$ with a polynomial P of degree 4 can be reduced to a similar integral with another polynomial of degree $\leqslant 3$ (hence of the type (3.24) at worst). We replace the integral by the differential equation satisfied by the inverse function and study thus

(3.39) $$(dx/dz)^2 = P(x) \quad , \text{ P polynomial of degree 4 .}$$

We look for meromorphic solutions of this equation, with a complex variable $z \in \mathbb{C}$. Let us see what happens when we make homographic changes of function. For $g \in SL_2(\mathbb{C})$, define

$$x' = x^g = g^{-1}(x) \ , \ x = g(x') = \frac{ax' + b}{cx' + d} \text{ if } g = \begin{pmatrix} a & b \\ c & d \end{pmatrix} .$$

We can then compute

$$dx/dz = (cx' + d)^{-2}(dx'/dz) = J(g,x')(dx'/dz)$$

with the notation introduced in (I.3) for the Jacobian $J(g,x')$ (applied here to <u>complex</u> unimodular matrices g). Then (3.39) gives the modified equation

$$J(g,x')^2(dx'/dz)^2 = (dx/dz)^2 = P(x) = P(g(x')) \quad ,$$

(3.40) $$(dx'/dz)^2 = J(g,x')^{-2}P(g(x')) \quad .$$

It is natural to introduce the notation

$$P^g(x) = J(g,x)^{-2}P(g(x)) = (cx + d)^4 P(\frac{ax + b}{cx + d})$$

for the new polynomial (of degree $\leqslant 4$) obtained. It is then an easy computation to check $P^{gg'} = (P^g)^{g'}$ using the chain rule for derivation $J(gg',x) = J(g,g'(x)) \cdot J(g',x)$:

$$P^{gg'}(x) = J(gg',x)^{-2} P(gg'(x)) = J(g',x)^{-2} P^g(g'(x))$$

and this is $(P^g)^{g'}(x)$ by definition. Thus, if we put

$$T_g(P) = P^{g^{-1}} \quad \text{for } P \text{ polynomial of degree } \leq 4$$
$$\text{(with complex coefficients)} \quad ,$$

we define a representation $g \longmapsto T_g$ of the group $SL_2(\mathbb{C})$ in the complex vector space of polynomials of degree less or equal to four. <u>This representation admits two classical invariants</u> described as follows. If we write the polynomial P in the form

(3.42) $$P(x) = a_0 + 4a_1x + 6a_2x^2 + 4a_3x^3 + a_4x^4 ,$$

we define

(3.43)
$$g_2(P) = a_0a_4 - 4a_1a_3 + 3a_2^2$$
$$g_3(P) = \begin{vmatrix} a_0 & a_1 & a_2 \\ a_1 & a_2 & a_3 \\ a_2 & a_3 & a_4 \end{vmatrix}$$

(These two quantities play an important role in the algebraic determination of the roots of the quartic $P(x) = 0$, as in Euler's method, cf. Burnside and Panton, Theory of equations, II p.113.) A direct verification of $g_i(P) = g_i(P^g)$ is rather tedious, and some gain is obtained by checking it for the generators

(3.44) $$\begin{pmatrix} 0 & -1 \\ 1 & 0 \end{pmatrix} , \quad \begin{pmatrix} t & 0 \\ 0 & t^{-1} \end{pmatrix} , \quad \begin{pmatrix} 1 & a \\ 0 & 1 \end{pmatrix} \quad (t \in \mathbb{C}^\times \text{ and } a \in \mathbb{C}).$$

If we write a_i^g for the coefficients of P^g (defined with the factors 4,6 as in (3.42)), we see easily that

$$g = \begin{pmatrix} 0 & -1 \\ 1 & 0 \end{pmatrix} \text{ implies } a_i^g = a_{4-i} \quad ,$$

so that P^g has the same coefficients as P in opposite order. This leaves obviously g_2 and g_3 invariant. In the second case,

$$g = \begin{pmatrix} t & 0 \\ 0 & t^{-1} \end{pmatrix} \text{ implies } a_i^g = t^{2i-4}a_i \quad ,$$

and again this substitution leaves g_2 and g_3 invariant. For the last type of generators, we show the invariance under the Lie algebra of the tangent subgroup. In other words, if $g = \begin{pmatrix} 1 & a \\ 0 & 1 \end{pmatrix}$ we show that the differentiable functions $g_2(P^g)$, $g_3(P^g)$ of a have a zero derivative (at the origin or everywhere). For this we write the Taylor expansion of $P^g(x) = P(x+a)$:

$$P(x+a) = P(a) + P'(a)x + P''(a)x^2/2 + \dots ,$$

and find

$$a_0^g = P(a) \ , \ a_1^g = P'(a)/4 \ , \ a_2^g = P''(a)/12 \ ,$$
$$a_3^g = P^{(3)}(a)/24 \ , \ a_4^g = P^{(4)}(a)/24 \ .$$

From this we get

$$g_2(P^g) = \frac{1}{24}(P(a)P^{(4)}(a) - P'(a)P^{(3)}(a) + \tfrac{1}{2}P''(a)^2) \ ,$$

whence

$$(d/da)g_2(P^g) = \frac{1}{24}(P'(a)P^{(4)}(a) - P''(a)P^{(3)}(a) - P'(a)P^{(4)}(a) + $$
$$+ P''(a)P^{(3)}(a)) = 0 \ .$$

The assiduous reader will check it for g_3 ! Then we say that the polynomial P^g is in normal form when it is given by

$$P^g(x) = 4x^3 + ax + b \ .$$

With the notations introduced in (3.42), this gives

$$a_4 = 0 \ , \ a_3 = 1 \ , \ a_2 = 0 \ , \ a_1 = a/4 \ , \ a_0 = b \ ,$$

and if we compute the invariants defined in (3.43) we get

$$g_2(P^g) = g_2(P) = -a \ , \ g_3(P^g) = \begin{vmatrix} b & a/4 & 0 \\ a/4 & 0 & 1 \\ 0 & 1 & 0 \end{vmatrix} = -b \ .$$

This shows that P^g is in normal form precisely when

$$P^g(x) = 4x^3 - g_2(P)x - g_3(P) \ .$$

The important point is that <u>if</u> P <u>has no multiple roots, it has a transform</u> P^g (for some unimodular complex matrix g) <u>in normal form and</u> $g_2(P)^3 - 27g_3(P)^2 \neq 0$. To see this, we first make a homography of the form $x \longmapsto 1/(x-a)$ which has the effect of reversing the coefficients of $P(x-a)$

$$(x-a)^4 P(1/(x-a)) = P(a)x^4 + \alpha P'(a)x^3 + \dots \ .$$

Thus if we choose a root a of P(x), the term in x^4 will disappear in the transformed polynomial (similarly, a double root will give a transformed polynomial of degree 2 and a triple root will give a polynomial of degree 1 : in particular, if P has a root of order $\geqslant 3$ $g_2(P) = g_3(P) = 0$). Combining the latter homography with a suitable linear transformation $x \longmapsto a'x + b'$ will bring the polynomial in normal form. Finally, we may consider substitutions of the form

$$z \longmapsto z' = \lambda z \ , \ x \longmapsto x' = \lambda^{-2}x \ .$$

They transform the initial differential equation , supposed in normal form, in

$$(dx'/dz')^2 = 4x'^3 - \lambda^{-4}g_2(P)x' - \lambda^{-6}g_3(P) \ ,$$

hence we get a new polynomial $\widetilde{P}$ with invariants

$$\widetilde{g}_2 = g_2(\widetilde{P}) = \lambda^{-4}g_2(P) \ , \ \widetilde{g}_3 = g_3(\widetilde{P}) = \lambda^{-6}g_3(P) \ .$$

This proves that if we can solve one differential equation (3.39), then we can solve all differential equations of the same type having the same invariant

$$j(P) = (12g_2)^3/(g_2^3 - 27g_3^2) = 1728 \ g_2^3/\Delta$$

(this invariant is finite if the differential equation cannot be solved by using the elementary trigonometric functions, i.e. when P has no multiple root). Differential equations (3.39) with same invariant j are essentially equivalent, and it is sufficient to solve one equation in each class , determined by a given $j \in \mathbb{C}$. Coming back to the integral form of the differential equation, it can be seen that the meromorphic solutions admit necessarily two independent periods (we refer to standard textbooks for this point, e.g. Valiron, Théorie des Fonctions, p.422).

4. Analytic p-adic functions

We gather here a few facts on p-adic numbers which will be used in later sections. Let p be a prime number and $\mathbb{Z}_p = \varprojlim_n \mathbb{Z}/p^n\mathbb{Z}$ be the ring of p-adic integers. This is a compact totally disconnected (additive) abelian group, and its elements admit unique representations

$$a = \sum_{n\geqslant 0} a_n p^n \quad \text{with} \quad 0 \leqslant a_n \leqslant p-1 \quad .$$

Addition and multiplication of two such series is done naturally (these laws are completely determined on the finite such series and extend uniquely by continuity, but note that they are not the formal series laws in the indeterminate p because for example, the sum of two a_n's with sum $\geqslant p$ will contribute a higher power of p). Because every integer $m \in \mathbb{Z}$ has a non-zero image in a $\mathbb{Z}/p^n\mathbb{Z}$ (for n sufficiently large) we deduce that there is a canonical embedding $\mathbb{Z} \longrightarrow \mathbb{Z}_p$ (integers $n \geqslant 0$ corresponding to polynomials in p), and $\mathbb{Z}_p$ is a ring of characteristic 0. It is easy to see that $\mathbb{Z}_p$ is an integral ring. We denote by $\mathbb{Q}_p$ its field of quotients so that we have two canonical embeddings $\mathbb{Z}_p \longrightarrow \mathbb{Q}_p$ and $\mathbb{Q} \longrightarrow \mathbb{Q}_p$. We identify $\mathbb{Z}_p$ and $\mathbb{Q}$ with their images in $\mathbb{Q}_p$. Elements of $\mathbb{Q}_p$ admit representations as Laurent series in p of the form

$$a = \sum_{n\geqslant n_0} a_n p^n \quad \text{with} \quad 0 \leqslant a_n \leqslant p-1 \quad .$$

If we denote by ord_p the order of the Laurent series : $\mathrm{ord}_p(a)$ is the smallest index in the Laurent series of a such that $a_n \neq 0$, the topology of $\mathbb{Q}_p$ is also given by the norm

$$|a| = |a|_p = p^{-\mathrm{ord}_p(a)} \in \mathbb{R}_+^\times \quad .$$

Then $\mathbb{Z}_p$ is the <u>closed unit ball</u> in $\mathbb{Q}_p$ defined by $|a| \leqslant 1$ (it is also <u>open</u> and defined by $|a| < p$). In particular, $\mathbb{Q}_p$ is <u>locally compact</u>.

We could also have defined equivalently $\mathbb{Q}_p$, resp. $\mathbb{Z}_p$, as completed of $\mathbb{Q}$, resp. $\mathbb{Z}$, for the topology defined by the p-adic absolute value $|a|_p = p^{-v_p(a)}$ where $v_p(a)$ is the power of p in the decomposition of a as product of prime powers (and a sign). One sees easily that

$$\mathbb{Q} \cap \mathbb{Z}_p = \mathbb{Z}_{(p)} = \{m/n : m,n \in \mathbb{Z} \text{ and } (n,p) = 1\}$$

is the localized of $\mathbb{Z}$ at the prime ideal (p) , and thus $\mathbb{Z}_p$ is also the completed of the local ring $\mathbb{Z}_{(p)}$ topologized by taking the powers $(p)^k = (p^k)$ of the maximal ideal $(p) = p\mathbb{Z}_{(p)}$ as basis of neighbourhoods of $0 \in \mathbb{Z}_{(p)}$.

Let K be any finite algebraic extension of $\mathbb{Q}_p$. The p-adic absolute value of $\mathbb{Q}_p$ has an extension to K (because the additive valuation $v_p = \mathrm{ord}_p$ has an extension) and in fact a unique one because K is a finite-dimensional normed space over the complete field $\mathbb{Q}_p$, and any two norms on such a vector space are equivalent. The existence and uniqueness of an extended absolute value lead to the following expression. Let $a \in K$ and L the normal extension generated by a (in an algebraic closure $\bar{\mathbb{Q}}_p$ of $\mathbb{Q}_p$). Then for every absolute value $|\ldots|$ on L and every automorphism σ of L over $\mathbb{Q}_p$, $x \longrightarrow |x^\sigma|$ is also an absolute value on L, so that $|x| = |x^\sigma|$ by uniqueness, and this shows that if we denote by $N = N_{L/\mathbb{Q}_p}$ the norm of L, and by $n = [L : \mathbb{Q}_p]$ the degree of L, then the extension of $|\ldots|$ to the element a is given by the explicit formula

$$|a| = |N(a)|^{1/n} \quad .$$

In particular, we see that the extended absolute value is invariant under all automorphisms over $\mathbb{Q}_p$, and this proves that these automorphisms are continuous for this norm. Because each element of $\bar{\mathbb{Q}}_p$ is contained in a finite extension of $\mathbb{Q}_p$, it follows that $\bar{\mathbb{Q}}_p$ itself is a valued field.

But note that $\bar{\mathbb{Q}}_p$ is of infinite degree over $\mathbb{Q}_p$: by Eisenstein's criterium, the polynomial $X^n - p$ is irreducible over $\mathbb{Q}_p$ hence any of its roots will give an extension of degree n (arbitrarily large) of $\mathbb{Q}_p$. Let π be any root of the preceeding equation. Its absolute value must satisfy $|\pi|^n = |p| = 1/p$, hence $|\pi| = p^{-1/n}$. If K denotes the field obtained by adjoining π to $\mathbb{Q}_p$: $K = \mathbb{Q}_p(\pi)$, this shows that the value group $|K^\times|$ in $\mathbb{R}_+^\times$ contains the subgroup

$$p^{(1/n)\mathbb{Z}} = \{p^{m/n} : m \in \mathbb{Z}\} \subset \mathbb{R}_+^\times .$$

Letting n tend to infinity, we conclude that

(4.1) $$|\bar{\mathbb{Q}}_p^\times| \supset p^{\mathbb{Q}} = \{p^{m/n} : m,n \in \mathbb{Z}\} \subset \mathbb{R}_+^\times$$

It would be easy to show that this value group is exactly $p^{\mathbb{Q}}$, but we only need to know that it is dense in $\mathbb{R}_+^\times$ (any non-empty open interval of $\mathbb{R}_+^\times$ contains infinitely many absolute values $|x|$ with $x \in \bar{\mathbb{Q}}_p^\times$). Being of infinite dimension over $\mathbb{Q}_p$, $\bar{\mathbb{Q}}_p$ is no more locally compact and not complete for the uniform structure determined by its absolute value (it is a metric space). We denote by Ω_p the completed field of $\bar{\mathbb{Q}}_p$. By construction (or definition) the absolute value of $\bar{\mathbb{Q}}_p$ has a unique extension to Ω_p (and $|\Omega_p^\times| = |\bar{\mathbb{Q}}_p^\times| \subset \mathbb{R}_+^\times$). It is important to know that Ω_p is algebraically closed, hence is a "universal field" (corresponding to $\mathbb{C}$ which is also complete and algebraically closed), and it will play the role of the complex field for a transcendental study of p-adic elliptic curves. This result follows from the general lemma :

(4.2) Lemma. Let K be an ultrametric valued field and $\widehat{K}$ its completed field (this is still an ultrametric valued field). If K is algebraically closed, so is $\widehat{K}$.

Proof. Let $P = P(X) = \sum a_i X^i$ be a non-constant unitary polynomial of $\widehat{K}[X]$ ($a_i \in \widehat{K}$), and look at the finite extension L of K defined by adding a root of P to $\widehat{K}$:

$$L = \widehat{K}[X]/(P) = \widehat{K}(\xi) = \widehat{K}[\xi] \quad ,$$

with the canonical image ξ of X in the quotient ring (a field). The absolute value of $\widehat{K}$ has a unique extension to L (again because L is a finite dimensional vector space over the complete valued field $\widehat{K}$). Now fix $\delta>0$ and take a unitary polynomial $\widetilde{P}\in K[X]$ of the same degree as P and with coefficients $\widetilde{a}_i \in K$ close to the corresponding coefficients a_i of P : $|\widetilde{a}_i - a_i| < \delta$. Because K is supposed to be algebraically closed, all its roots ξ_i lie in K and we can write

$$\widetilde{P}(X) = \prod (X - \xi_i) \qquad (\xi_i \in K) \; .$$

Then on one hand $\widetilde{P}(\xi) = \prod (\xi - \xi_i)$ and on the other

$$\widetilde{P}(\xi) = (\widetilde{P} - P)(\xi) = \sum (\widetilde{a}_i - a_i)\xi^i \quad ,$$

so that we get the following estimate for the absolute values

$$\prod |\xi - \xi_i| \leqslant \operatorname*{Max}_i |\widetilde{a}_i - a_i| \cdot |\xi|^i < \delta \cdot \operatorname*{Max}_i |\xi|^i = \delta \cdot M_\xi \quad .$$

This implies that for one i at least

$$|\xi - \xi_i| < (\delta M_\xi)^{1/n} < \varepsilon \qquad (\xi_i \in K) \; ,$$

as soon as δ is chosen according to $\delta < \varepsilon^n / M_\xi$. The preceding argument shows that ξ belongs to the closure of K : $\xi \in \widehat{K}$. Consequently $\widehat{K}$ is algebraically closed.

As we shall use several ultrametric properties of Ω_p , we recall some of them briefly here. An <u>ultrametric space</u> in general, is a metric space equipped with a distance d satisfying the stronger inequality

$$d(x,z) \leqslant \operatorname{Max}(d(x,y),d(y,z)) \quad \text{for all } x,y,z \; .$$

Let us introduce the following convenient notations and terminology. The closed ball $B'_r(a)$ of radius r and center a is defined by $d(x,a) \leqslant r$. The open ball $B_r(a)$ of radius r and center a is defined by $d(x,a) < r$. The sphere $U_r(a)$ is defined by $d(x,a) = r$. (Note the perversity of the language, because all these sets are both open and closed in an ultrametric space, as will follow from the next observations !) If $b \in B_r(a)$, then $B_r(b) = B_r(a)$ because

$$d(x,a) < r \text{ and } d(a,b) < r \text{ imply } d(x,b) < r ,$$

and interchange a and b. Thus every point of $B_r(a)$ is a possible center. A similar result holds for closed balls. If $x \notin B_r(a)$, then the distance d(x,a) is equal to the distance of x to any point $b \in B_r(a)$: $d(x,b) = d(x,a)$ can be thought of as the distance of x and the ball $B_r(a)$ (again, we can replace $B_r(a)$ by $B'_r(a)$ if $x \notin B'_r(a)$). Consequently two disjoint balls $B_r(a)$ and $B_s(b)$ are at a well defined distance from each other :

$$d(x,y) = d(a,b) \quad \text{is independent from } x \in B_r(a), y \in B_s(b) .$$

If two balls have a common point, the smallest one is contained in the biggest one, and in particular, two balls of the same radius having a non-empty intersection coïncide. We note also that if $d(x,a) = r$ and $s < r$, then $B'_s(x)$ is contained in the sphere of radius r : $B'_s(b) \subset U_r(a)$. These spheres are thus open sets and $B'_r(a)$ being the union $B_r(a) \cup U_r(a)$ is open. Similarly $B_r(a) = B'_r(a) - U_r(a)$ is closed. The situation concerning Cauchy sequences is also particularly simple : a sequence (x_n) in an ultrametric space is a Cauchy sequence exactly when $d(x_n, x_{n+1}) \longrightarrow 0$ (a dream of youth), and in a commutative complete ultrametric group, a family (x_n) is absolutely summable exactly when $x_n \longrightarrow 0$.

These properties are in particular true for the complete algebraically closed valued field Ω_p , and they enable to make a very simple theory of analytic functions on Ω_p . The Cauchy integral cannot be defined in this context, and all proofs have to be derived by direct reasoning on power sequences (the handling of which is fortunately much simpler than in the complex case).

First we prove a p-adic analogue of the classical theorem of Rouché (for polynomials).

(4.3) Theorem 1. Let K be a finite algebraic extension of $\mathbb{Q}_p$, $f,g \in K[X]$ be two polynomials with $f(0) = g(0) = 1$. Suppose that f has n (≥ 1) roots on the unit sphere $U = U_1(0)$ of $\bar{K} = \bar{\mathbb{Q}}_p$ (or of Ω_p), and that $|f(x) - g(x)| < \varepsilon^n$ for all $x \in \bar{K}$ with $|x| \leq 1$ (i.e. $x \in B_1'(0)$), where $0 < \varepsilon < 1$. Then g has same number of roots as f on U and they are moreover equally partitioned among the closed balls $B_\varepsilon'(a)$ of radii ε contained in U ($a \in U$).

Proof. Let us number the roots a_i of f by taking those on U first. Then decompose f as product of linear terms

$$f(X) = \prod_{i=1}^{n} (1 - X/a_i) \prod_{|a_i| \neq 1} (1 - X/a_i) ,$$

and similarly for g

$$g(X) = \prod_{j=1}^{m} (1 - X/b_j) \prod_{|b_j| \neq 1} (1 - X/b_j) .$$

Now observe that for x on the unit sphere U, i.e. $|x| = 1$, we have

$$|1 - x/a_i| = \begin{cases} 1 \text{ if } |a_i| > 1 \\ |x/a_i| = |a_i|^{-1} > 1 \text{ if } a_i < 1 \end{cases} .$$

Thus for $x \in U$ one has

$$|f(x)| = A \prod_{i=1}^{n} |1 - x/a_i| ,$$

$$|g(x)| = B \prod_{j=1}^{m} |1 - x/b_j| ,$$

with two constants $A \geq 1$, $B \geq 1$. We now look at the balls $B_\varepsilon'(a_i)$ (disjoint if distinct) centered at the roots of f in U ($1 \leq i \leq n$) and choose $\delta > \varepsilon$ close to ε so that the balls $B_\delta(a_i)$ are distinct (or disjoint) at the same time as the corresponding smaller $B_\varepsilon'(a_i)$. We let x vary in the region (annulus)

$$D_1 = B_\delta(a_1) - B_\varepsilon'(a_1) \subset U \ (\subset \bar{K} \text{ or } \Omega_p) ,$$

defined equivalently by $\varepsilon < |x - a_1| < \delta$. Because $B_\delta(a_i)$ is disjoint from $B_\delta(a_1)$ if $B'_\varepsilon(a_i) \neq B'_\varepsilon(a_1)$, we infer that $|x - a_i| > \delta > \varepsilon$ for those i and we get

$$|f(x)| = A \prod_{i=1}^{n} |x - a_i| > A\varepsilon^n > \varepsilon^n \quad \text{for } x \in D_1 .$$

But $|f(x) - g(x)| < \varepsilon^n$ for the same x, hence by the ultrametric property, necessarily $|f(x)| = |g(x)|$ for $x \in D_1$. More precisely, still for $x \in D_1$, if we denote by n_1 the number of roots of f in $B'_\varepsilon(a_1)$,

$$\prod_{i=1}^{n} |x - a_i| = \prod_{a_i \in B'_\varepsilon(a_1)} \underbrace{|x - a_i|}_{=|x - a_1|} \prod{}' \underbrace{|x - a_i|}_{=|a_i - a_1|} .$$

This shows that $|f(x)| = A_1 |x - a_1|^{n_1}$ for $x \in D_1$, and similarly if m_1 denotes the number of roots of g in $B'_\varepsilon(a_1)$, we get $|g(x)| = B_1 |x - a_1|^{m_1}$ for $x \in D_1$ (note that $|x - b_j| = |x - a_1|$ if $b_j \in B'_\varepsilon(a_1)$). Since we know that f and g have same modulus on D_1 , we get

$$|x - a_1|^{n_1 - m_1} \quad \text{independent of } x \text{ in } D_1 .$$

Taking $x, x' \in D_1$ with distinct distances in $]\varepsilon, \delta[$ from a_1 (this is possible because the set of absolute values is dense in $]\varepsilon, \delta[$) proves $n_1 = m_1$. This shows more generally that f and g have same number of roots in all closed balls $B'_\varepsilon(a_i)$. For any other ball B'_ε contained in U, the inequality $|f(x)| > \varepsilon^n$ will still be valid in B'_ε hence $|g(x)| = |f(x)|$ in B'_ε and neither f nor g have roots in B'_ε . This proves the statement completely.

We can now study power series on Ω_p . Let $f \in \Omega_p[[X]]$ be a formal series

$$f(X) = \sum_{n \geq 0} a_n X^n \qquad (a_n \in \Omega_p) .$$

If $\sum a_n x^n$ converges for some $x \in \Omega_p^\times$, this means $a_n x^n \to 0$, or $|a_n| |x|^n \to 0$ and consequently $|a_n| |y|^n \to 0$ for every $y \in \Omega_p$ in the closed ball $|y| \leqslant |x|$ and the series will converge normally on that closed ball (absolutely at each point and uniformly on the ball).

(This attractive situation is well-known to fail with ordinary complex power series!) Under these circumstances, f defines a mapping $B'_{|x|}(0) \longrightarrow \Omega_p$. If all coefficients a_n lie in a finite sub-extension K of $\mathbb{Q}_p$, and if y is also in K with $|y| \leqslant |x|$, the value f(y) will also be in K because this field is complete (and all partial sums are in K). The whole theory is based on the fact that the modulus $|f(x)|$ can be given more or less explicitely, at least as far as x is not on certain exceptional spheres. We have to remember that, in a finite (or infinite) sum, a term alone of maximal size (absolute value) carries all responsability for the modulus of the whole sum ! For instance, if $a_0 \neq 0$, necessarily $|f(x)| = |a_0|$ in a neighbourhood (an open disc) of x = 0. More generally, if n is the order of $f \neq 0$ at the origin, i.e. the smallest index i with $a_i \neq 0$, we shall have $|f(x)| = |a_n||x|^n$ in a neighbourhood of x = 0. If however $f(X) \neq a_n X^n$, the size of a term $a_m x^m$ $(m > n)$ with $a_m \neq 0$ will necessarily overtake the size of $a_n x^n$ for $|x|$ large enough. This leads to the notion of critical radius.

(4.4) <u>Definition</u>. <u>Suppose that the power series defined by</u> f <u>converges in an open ball</u> $B_R = B_R(0)$ <u>with some positive</u> $R > 0$. <u>Then a radius</u> $0 < r < R$ <u>is called critical radius for</u> f <u>if there are two distinct indices</u> $0 \leqslant m < n$ (<u>at least</u>) <u>with</u>

$$|a_m| r^m = |a_n| r^n = \operatorname*{Max}_i (|a_i| r^i) \quad .$$

Obviously, a critical radius occurs in $|\Omega_p^\times|$ because with the notations of the definition,

$$r = |a_m/a_n|^{1/(n-m)} \quad .$$

Critical radii can be ordered in increasing sequence $0 < r_1 < r_2 < \dots$ There is a very useful notion connected with critical radii.

(4.5) <u>Definition</u>. <u>Let</u> r <u>be a critical radius for the power series defined by</u> f. <u>Then the corresponding critical index</u> n_r <u>is by</u>

<u>definition the biggest index</u> i <u>with</u> $|a_i| r^i = \operatorname{Max}_j |a_j| r^j$.

If $r_1, r_2, \ldots$ denotes the increasing sequence of critical radii of f, we denote by $n_1, n_2, \ldots$ the corresponding (increasing) sequence of critical indices : $n_i = n_{r_i}$. By the remarks preceeding the definition, it is obvious that

(4.6) $$|f(x)| = |a_{n_i}| r^{n_i} \quad \text{for} \quad r_i < |x| = r < r_{i+1} ,$$

only depends on the modulus of x in the annulus between two adjacent critical (exceptional) spheres. On a critical sphere, we can only assert an inequality

(4.7) $$|f(x)| \leqslant |a_{n_i}| r_i^{n_i} \quad \text{for} \quad |x| = r_i .$$

This has as obvious consequence the following proposition.

(4.8) <u>Proposition</u>. <u>If</u> f <u>is given by a convergent power series (in some open disc) with</u> $f(0) \neq 0$, <u>then the zeros of</u> f <u>can only occur on the critical spheres (having a critical radius)</u>.

More striking will be the result that f has indeed zeros on <u>all</u> critical spheres (and in fact only finitely many on each sphere). The p-adic analogue of Liouville's theorem is also a direct consequence of (4.6). An <u>entire function</u> $f : \Omega_p \longrightarrow \Omega_p$ is a function given by a power series which converges at all points $x \in \Omega_p$. Thus entire functions are defined by those power series having coefficients a_n satisfying $|a_n| = o(r^{-n})$ for every positive number r.

(4.9) <u>Proposition</u>. <u>If</u> $f : \Omega_p \longrightarrow \Omega_p$ <u>is an entire function with some majoration of the form</u> $|f(x)| \leqslant M|x|^N$ <u>for all</u> $x \in \Omega_p$ (<u>with a positive constant</u> M <u>and a positive integer</u> N), <u>then</u> f <u>is a polynomial of degree smaller or equal to</u> N.

<u>Proof</u>. If $|f(x)| \leqslant M|x|^N$ is true throughout Ω_p , then (4.6) shows that there is no critical index $n_i > N$. This implies that all a_n with index $n > n_i$ vanish and proves the proposition.

We have to consider a slightly more general situation, namely that of Laurent series. Let thus $f(X) = \sum_{-\infty}^{\infty} c_i X^i$ be a formal series with coefficients $c_i = c_i(f)$ in a fixed finite extension K of $\mathbb{Q}_p$ (the reader will observe that K could be replaced by the universal domain Ω_p if we did not want to prove rationality conditions). We suppose that f(x) is convergent in a non-empty open annulus and we denote by

$$\dots < r_{-1} < r_0 < r_1 < \dots$$

the sequence of critical radii of f (this sequence could be empty, but if f converges for all $x \in \Omega_p$, it can be empty only if f is reduced to a monomial : more precisely, if there are infinitely many non-zero coefficients c_i with positive indices i, the sequence of critical radii will be unlimited at the right, whereas if there are infinitely non-zero coefficients c_i with negative indices, the sequence will be unlimited at the left). Let also

$$\dots < n_{-1} < n_0 < n_1 < \dots$$

be the corresponding sequence of critical indices. By definition, n_i is the biggest integer n with $|a_n| r_i^n$ maximal. Similarly, if we define m_i to be the smallest integer n with $|a_n| r_i^n$ maximal, we have obviously $m_i = n_{i-1}$. We shall fix our attention to one particular critical radius which we take to be $r_0 = 1$ (changing of variable X if necessary). Multiplying f by a suitable monomial we may assume

$$(4.10) \quad \begin{array}{l} m_0 = n_{-1} = 0 \ , \ n_0 = n > 0 \\ c_0 = 1 \ , \ |c_n| = 1 \ , \ |c_i| \leqslant 1 \ \text{(strict inequality for } i < 0 \text{ and for } i > n) \ . \end{array}$$

We have then

$$(4.11) \qquad |f(x)| = \begin{cases} 1 & \text{if } r_{-1} < |x| < 1 \\ |x|^n & \text{if } 1 < |x| < r_1 \end{cases} \quad .$$

To fix ideas, we suppose that there are infinitely many coefficients $c_i \neq 0$ with negative indices i (otherwise we would change X to X^{-1}, the case of a polynomial in X and X^{-1} would be trivial in the considerations below), and we take truncated sums

$$f_M^N(X) = \sum_{M \leqslant i \leqslant N} c_i X^i \qquad \text{with } M \leqslant 0 < n \leqslant N \text{ , } c_M \neq 0 \text{ .}$$

Let us introduce the roots $a_i = a_i(M,N)$ of the polynomial $f_M^N/c_M X^M$:

(4.12) $$f_M^N(X) = c_M X^M \prod (1 - X/a_i) \text{ .}$$

Because the coefficients of f_M^N are the same as those of f for $0 \leqslant i \leqslant n$, the critical radii r_i' of f_M^N satisfy $r'_{-1} = r_{-1}$, $r'_0 = r_0$, $r'_1 = r_1$, and in particular, the roots a_i not on the unit sphere, satisfy

(4.13) $$\text{either } |a_i| \leqslant r_{-1} \text{ or } |a_i| \geqslant r_1 \text{ .}$$

With (4.12) we can give the following expressions for the modulus of the function f :

(4.14) $$|f_M^N(x)| = \begin{cases} |c_M||x|^M \prod_{|a_i|<1} |x/a_i| & \text{if } r_{-1} < |x| < 1 \\ |c_M||x|^M \prod_{|a_i|\leqslant 1} |x/a_i| & \text{if } 1 < |x| < r_1 \text{ .} \end{cases}$$

But in these two regions, the leading terms are $c_0 = 1$ and c_n respectively (as for f) and (4.11) is valid for f_M^N :

$$|f_M^N(x)| = \begin{cases} 1 & \text{if } r_{-1} < |x| < 1 \\ |x|^n & \text{if } 1 < |x| < r_1 \text{ .} \end{cases}$$

Comparing these two estimates, we get immediately the following :

$$-M = |M| = \text{number of roots } |a_i| < 1 \text{ ,}$$

(4.15) $$|c_M| = \prod_{|a_i|<1} |a_i| < 1 \text{ ,}$$

$$n = \text{number of roots } |a_i| = 1 \text{ (independent of M and N) .}$$

Now we let $-M = |M|$ and N tend to infinity and consider the variation of the roots $a_i(M,N)$. I claim that if they are properly numbered (considering i as fixed) they make up Cauchy sequences. Take indeed another bigger truncated $f_{M'}^{N'}$ with $M' < M$ and $N' > N$, and suppose that $|f_M^N(x) - f_{M'}^{N'}(x)| < \varepsilon^n$ for $|x| = 1$. Call simply a_i' the roots of the polynomial corresponding to $f_{M'}^{N'}$. Using (4.15) for this new truncated, together with (4.12), we get

$$\left|f_{M'}^{N'}(x)\right| = \prod_{1\leqslant j\leqslant n} \left|1 - x/a_j'\right| = \prod_j |x - a_j'| \quad \text{for } |x| = 1 \quad .$$

Substituting the root a_i of f_M^N for x, we get

$$\prod_{1\leqslant j\leqslant n} |a_i - a_j'| = \left|f_{M'}^{N'}(a_i) - f_M^N(a_i)\right| < \varepsilon^n ,$$

hence for one j at least $|a_i - a_j'| < \varepsilon$. Because the roots $a_i \in \bar{K}$, the limit roots $a_i^{\infty} = \lim a_i(M,N) \in \Omega_p$. Let us introduce a notation for the normalized polynomials having these roots :

$$g_M^N(X) = \prod (1 - X/a_i(M,N)) \ , \ g(X) = \prod (1 - X/a_i^{\infty}) \ .$$

These are polynomials of degree n.

<u>First step</u>. <u>The limit roots</u> a_i^{∞} <u>are roots of</u> f. Indeed we have

$$\left|f(a_i(M,N))\right| = \left|(f - f_M^N)(a_i(M,N))\right| \leqslant \operatorname*{Max}_{\substack{i\leqslant M\\ i\geqslant N}} |c_i(f)| \ .$$

Letting $M\to -\infty$, and $N\to\infty$, and using the continuity of f (remember that $|c_i|\to 0$ for $|i|\to\infty$) we get $f(a_i) = 0$.

<u>Second step</u>. For every automorphism $\sigma \in \mathrm{Gal}(\bar{K}/K)$ and every root a_i of f_M^N , a_i^{σ} is still a root of f_M^N (which has coefficients in K) and moreover $|a_i^{\sigma}| = |a_i|$. This proves that the set of roots a_i on the unit sphere is invariant under σ , and by Galois theory, $g_M^N \in K[X]$ will have coefficients in K . Because the roots of g_M^N make up Cauchy sequences, the coefficients of g_M^N (which are symmetric combinations of roots) will also make up Cauchy sequences in the complete field K. This proves $g \in K[X]$ and consequently <u>the limit roots</u> a_i^{∞} (roots of g) <u>are in the algebraic closure</u> $\bar{K}$ <u>of</u> K.

<u>Third step</u>. <u>Then</u> f = g·h <u>where</u> $h(X) = \sum_{-\infty}^{\infty} c_i(h)X^i$ <u>has no root on the unit sphere</u> (r = 1 <u>is no more a critical radius for</u> h) <u>has coefficients</u> $c_i(h) \in K$ <u>and converges for all</u> $x \in \Omega_p^{\times}$. Write $f_M^N = g_M^N \cdot h_M^N$. Because $g_M^N(X) = 1 + \ldots \pm \left(\prod a_i\right)^{-1} X^n$ has all its coefficients in the ring of integers R_K of K defined by $|x| \leqslant 1$ (and so does f),

the coefficients $c_i(h_M^N) \in R_K$ will also be integral (Gauss' lemma). For each fixed i, the coefficients $c_i(f_M^N)$ and $c_i(g_M^N)$ converge in R_K. Thus the coefficients $c_i(h_M^N)$ converge in R_K, and this proves the existence of the formal series h, with integral coefficients. Using reduction mod the maximal ideal M_K of R_K defined by $|x|<1$, we get the equality $\tilde{f} = \tilde{g}\cdot\tilde{h}$. But $\tilde{f}$ and $\tilde{g}$ are polynomials of degree n and constant term unity. Hence $\tilde{h} = 1$ and h has only one coefficient of absolute value one : r = 1 is not a critical radius for h.
In fact, more precisely, we have the inequalities

$$|c_i(h)| \leqslant \sup_{j\geqslant n} |c_{i+j}(f)| \qquad i \geqslant 0$$

$$|c_i(h)| \leqslant |c_i(f)| \qquad i \leqslant 0 \quad ,$$

because one can check these inequalities for the truncated polynomials (g_M^N has two extreme coefficients of unit modulus) and they will remain valid at the limit. We can then state the main result of this section.

(4.16) <u>Theorem</u> (<u>L. Schnirelmann</u>). <u>Let</u> $f(X) = \sum_{-\infty}^{\infty} c_i X^i$ <u>be a formal Laurent series with coefficients</u> c_i <u>in a finite extension</u> K <u>of</u> $\mathbb{Q}_p$. <u>We suppose that</u> f(x) <u>converges for all</u> $x \in \Omega_p^\times$. <u>Then</u>, f(X) <u>can be written in the form</u>

$$f(X) = cX^k \prod_{|\alpha|<1} (1 - \alpha/X) \prod_{|\alpha|\geqslant 1} (1 - X/\alpha)$$

<u>with finite non-empty sets of roots</u> $\alpha \in \bar{K}$ <u>occuring on the critical spheres of</u> f. <u>Gathering these roots of given modulus together, we get a representation</u>

$$f(X) = cX^k \prod_{i<0} \hat{g}_i(X) \prod_{i\geqslant 0} g_i(X) \quad ,$$

<u>with polynomials</u> $g_i(X) \in K[X]$ <u>or</u> $\hat{g}_i(X) \in K[X^{-1}]$ <u>having the same roots as</u> f <u>on the critical spheres of radii</u> r_i, $c \in K$, $k \in \mathbb{Z}$.

<u>Proof</u>. As has been seen, we can remove the critical radii of f one after the other, by dividing by polynomials (in X or X^{-1}).

Then we observe that the infinite products

$$\prod_{|\alpha|<0} (1 - \alpha/X) = \prod_{i<0} \hat{g}_i(X) \ , \quad \prod_{|\alpha|>0} (1 - X/\alpha) = \prod_{i>0} g_i(X) \ ,$$

are convergent for all $x \in \Omega_p^\times$, and hence, going to the limit, we can write $f(X) = h(X) \prod_{i<0} \hat{g}_i(X) \prod_{i>0} g_i(X)$ with a convergent power series h having no critical radius left. This shows that h(X) is a monomial of the form cX^k with a certain rational integer $k \in \mathbb{Z}$ and a constant $c \in K$. q.e.d.

If f is as above, and L is any finite extension of K, then for every $x \in L^\times$, f(x) will belong to the complete field L. Because $\bar{K}$ is the union of such finite (complete) subextensions L we get

(4.17) $$f(\bar{K}^\times) \subset \bar{K} \ .$$

But the theorem shows that conversely if $a \in \bar{K}$ the solutions of $f(x) = a$ are also in $\bar{K}$: consider $L = K(a)$ and the formal series $f - a$ which has all its coefficients in L. All its zeros are in $\bar{K}$.

(4.18) Corollary 1. If f is as above, f determines mappings

$$f : K^\times \longrightarrow K \ , \ \bar{K}^\times \longrightarrow \bar{K} \ , \ \Omega_p^\times \longrightarrow \Omega_p \ .$$

Then the inverse image $f^{-1}(\bar{K})$ of $\bar{K}$ in $\Omega_p^\times$ is contained in $\bar{K}$ (f can take algebraic values only at algebraic points).

(4.19) Corollary 2. If f has an essential singularity at the origin (infinitely many Laurent coefficients $c_i(f) \neq 0$ for $i < 0$), then f takes all values $a \in \bar{K}$ (infinitely many times) in all punctured balls $B_\varepsilon(0) - \{0\}$ of $\bar{K}$ (where $\varepsilon > 0$ is any positive number) (p-adic analogue of the big Picard theorem).

Proof. With the assumptions of this corollary, f has an infinite sequence $(r_i)_{i<0}$ of critical radii with $r_i \longrightarrow 0$ for $i \longrightarrow -\infty$, and on each radius r_i , f has a zero in $\bar{K}$. Hence the value 0 is taken infinitely many times in all punctured neighbourhoods of 0. Replacing f by $f - a$ (where $a \in \bar{K}$ is arbitrary) gives the announced result.

(4.20) Corollary 3. Let $f(X)$ be a formal Taylor series $f(X) = \sum_{i \geq 0} c_i X^i$ with coefficients $c_i \in K$, $f(0) = c_0 \neq 0$, converging in an open ball $B_r(0) \subset \Omega_p$. If f does not vanish in this ball, the formal power series for $1/f$ will also converge in $B_r(0)$.

Proof. By the above theorem, f has no critical radius in $|x| < r$. Hence, if we assume $c_0 = 1$, $|f(x)| = 1$ for $|x| < r$ and

$$|c_i| \, t^i < 1 \text{ for } t < r .$$

But the inequalities $|c_i| \leq r^{-i}$ determine a subgroup of $1 + X \cdot K[[X]]$, as one checks immediately, using the ultrametric property of the absolute value.

For reference we formulate explicitely the following

(4.21) Corollary 4. If $f(X)$ is a formal Laurent series converging in the whole of $\Omega_p^\times$, and if f has no zero, then $f(X) = cX^k$ is a monomial in X.

In particular, there is no holomorphic function in the whole of Ω_p playing the role of the exponential of the classical function theory. In fact it is possible to define an exponential

$$e^X = \sum_{i \geq 0} X^i/i!$$

with the formal property $e^{X+Y} = e^X \cdot e^Y$. As consequence, e^x is inversible (with inverse e^{-x}) for every element $x \in \Omega_p$ in the domain of convergence of the series. This proves that the exponential series has no zero, no critical radius (and in particular cannot converge in the whole of Ω_p).

We introduce the following notations

(4.22)

H_K : ring of p-adic holomorphic functions on $\Omega_p^\times$ defined by convergent Laurent series with coefficients in K,

M_K : field of p-adic meromorphic functions on $\Omega_p^\times$ defined over K (field of fractions of H_K).

By Schnirelmann's theorem, a meromorphic function $f \in M_K$ has only finitely many zeros and poles in a closed annulus

$$0 < r \leq |x| \leq r' < \infty$$

and their multiplicities are well defined. Still by the same theorem, we see that if f has a zero of order d_a at the algebraic element $a \in \bar{K}^{\times}$, then it will have zeros of the same orders at all conjugates a^{σ} of a over K (and similarly for poles). Conversely, given any set of elements $a \in \bar{K}^{\times}$ with multiplicities $d_a \in \mathbb{Z}$ satisfying the preceding two conditions, it is possible to construct a canonical convergent Weierstrass product, defining a meromorphic function $f \in M_K$ having precisely the zeros (or poles) $a \in \bar{K}$ with respective multiplicities d_a. These data determine f up to a multiplicative factor of the form cX^k ($c \in K$, $k \in \mathbb{Z}$). This is what we shall use systematically in the next section.

(4.23) Remark. Only for simplicity have we assumed that the characteristic of K is 0, in our study of p-adic analytic functions. The results are true in general (i.e. also for formal power series fields $\mathbb{F}_q((t))$ over finite fields of characteristic p) as one checks easily. Only at one point does one have to be a little bit more careful in the proof of Schnirelmann's theorem, namely in the second step on p.II.81. The Galois invariance property only implies that the coefficients of g_M^N are purely inseparable over K. But if a root α of f_M^N has order of inseparability p^r over K (i.e. if r is the smallest integer such that α^{p^r} is separable over K), its multiplicity will be a multiple mp^r of this order in both f_M^N and g_M^N. This shows that the coefficients of g_M^N will be separable over K. Hence $g_M^N \in K[X]$ as in characteristic 0.

5. Tate's p-adic elliptic curves

In a first attempt to define p-adic elliptic curves, one is led to consider quotients Ω_p/L (instead of $\mathbb{C}/L$) with discrete subgroups L in the additive group of Ω_p (or in a finite algebraic extension K of $\mathbb{Q}_p$). However, this theory is soon recognized to collapse through lack of candidates! If $\omega \in L$ (a subgroup of a p-adic field), then

$$p^n \omega = \omega + \omega + \omega + \dots \in L \quad ,$$

and

$$|p^n\omega|_p = |\omega|/p^n \longrightarrow 0 \quad (n \to \infty)$$

implies $\omega = 0$ if L is discrete. Thus, the only discrete additive subgroup of a p-adic field is the trivial one $L = \{0\}$.

In the classical case, the normalized exponential $\underline{e} : \mathbb{C} \longrightarrow \mathbb{C}^\times$ transforms a lattice $L = \mathbb{Z} + \tau\mathbb{Z}$ $(\mathrm{Im}(\tau) > 0)$ into the multiplicative subgroup $q^{\mathbb{Z}} = \underline{e}(\tau\mathbb{Z})$ (with $q = \underline{e}(\tau) \in \mathbb{C}^\times$, $|q| < 1$). (Strictly speaking we should define $q^2 = \underline{e}(\tau)$ to keep perfect coherence with the first chapter. However, this would lead to using throughout even powers of q, and the reader will easily convince himself that this new normalization is more natural.) But now, there are plenty of discrete subgroups (of rank one) in $\Omega_p^\times$ (or in $K^\times$, with K as above). In fact, each $0 \neq |q| < 1$ generates such a subgroup $q^{\mathbb{Z}}$, and we shall concentrate our attention to the quotient $\Omega_p^\times/q^{\mathbb{Z}}$ (or to $K^\times/q^{\mathbb{Z}}$). This means that we shall work with Jacobi's point of view viz. p-adic theta functions.

Complex case	p-adic case
$\mathbb{C}/L$	no p-adic analogue
$\underline{e}\ \downarrow$ normalized exp	no exponential available
$\mathbb{C}^\times/q^{\mathbb{Z}}$	$\Omega_p^\times/q^{\mathbb{Z}}$: p-adic elliptic curve.

We repeat the definitions

(5.1) H_K : space of holomorphic functions on $\Omega_p^\times$ with coefficients $c_n \in K$ ($f(X) = \sum c_n X^n$ converging for $0 < |x| < \infty$) ,

M_K : space of meromorphic functions on $\Omega_p^\times$ given by quotient of two functions $f, g \neq 0$ in H_K .

Thus, by section 4, a meromorphic function $f/g \in M_K$ has a canonical Weierstrass expansion

$$f/g(X) = cX^k \prod_{|a|\leqslant 1} (1 - a/X)^{d_a} \prod_{|a|>1} (1 - X/a)^{d_a},$$

with rational integers $d_a \in \mathbb{Z}$ (for $a \in \bar{K}$) satisfying

(5.2) a) only finitely many multiplicities d_a are not 0 in any annulus of the form $0 < r' \leqslant |a| \leqslant r'' < \infty$,

b) for all $a \in \bar{K}^\times$ and $\sigma \in \mathrm{Gal}(\bar{K}/K)$, $d_a = d_{a^\sigma}$.

Then for $q \in K^\times$ and $|q| < 1$, we define the subfield $L_K = L_K^q$ of M_K as being the field of functions satisfying $f(qX) = f(X)$:

(5.3) $L_K = L_K^q = L_K(q)$: $f \in M_K$ and $f(q^{-1}X) = f(X)$ $(X \in \bar{K}^\times)$.

This field will be shown to be an elliptic function field over K. Every function $f \in L_K^q$ (we shall say q-elliptic function or simply elliptic function) has a well-defined divisor over $\bar{K}^\times/q^{\mathbb{Z}}$ satisfying the condition (5.2.b) above. The group of all such divisors will be denoted by

(5.4) $\mathrm{Div}_K(\bar{K}^\times/q^{\mathbb{Z}}) \subset \mathrm{Div}(\bar{K}^\times/q^{\mathbb{Z}})$,

and the divisors of this subgroup (satisfying (5.2.b)) will be called rational over K . Purely formally first, we put

(5.5) $E_q = E_q(\bar{K}) = \bar{K}^\times/q^{\mathbb{Z}}$.

We want to show that this set can be identified with the set of ($\bar{K}$ -) points on a non-singular cubic (elliptic curve) defined over K .

As we have already pointed out, the whole study is based on theta functions.

(5.6) Definition. A theta function θ (relative to the subgroup $q^{\mathbb{Z}}$) is a meromorphic function $\theta \in M_K$ on $\bar{K}^{\times}$ having a q-periodic divisor

$$\operatorname{div}\theta = \underline{d} \in \operatorname{Div}_K(E_q)$$

(this means that the set of zeros and poles of θ, together with the multiplicities, is invariant under $q^{\mathbb{Z}}$, and thus defines a divisor on E_q). If θ is a theta function, and $\theta'(X) = \theta(q^{-1}X)$, then θ' is also a theta function, and $\operatorname{div}\theta' = \operatorname{div}\theta$. Consequently by Schnirrelmann's theorem, $\theta'(X) = c^{-1}(-X)^d\theta(X)$, which means

$$\theta(q^{-1}X) = c^{-1}(-X)^d\theta(X) \qquad (c \in K^{\times},\ d \in \mathbb{Z}) .$$

If we define $\theta''(X) = c'X^k\theta(X)$, we see immediately that

$$\theta''(q^{-1}X) = (cq^k)^{-1}(-X)^d\theta''(X) \quad ,$$

and since θ'' is the most general expression of a theta function of divisor equal to that of θ, this proves that the class of c in $K^{\times}/q^{\mathbb{Z}}$ is well defined by the divisor $\operatorname{div}\theta = \underline{d}$, independently of the theta function chosen with this divisor. Because every divisor $\underline{d} \in \operatorname{Div}_K(E_q)$ is the divisor of a theta function, we can define two homomorphisms

(5.7)
$$\begin{aligned} d_q &: \operatorname{Div}_K(E_q) \longrightarrow \mathbb{Z} \quad ,\\ &\underline{d} \longmapsto d_q(\underline{d}) = d \quad ,\\ \phi_q &: \operatorname{Div}_K(E_q) \longrightarrow K^{\times}/q^{\mathbb{Z}} \quad ,\\ &\underline{d} \longmapsto \phi_q(\underline{d}) = c \bmod q^{\mathbb{Z}} \quad . \end{aligned}$$

(5.8) Proposition. The mapping d_q is the degree homomorphism, and ϕ_q is the Abel-Jacobi homomorphism which associates to the divisor $\underline{d} = \sum d_a(a)$, the element $c = \prod a^{d_a}$ in $K^{\times}/q^{\mathbb{Z}}$ (if the group law on $K^{\times}/q^{\mathbb{Z}}$ was noted additively, this element c would be Abel's sum $\sum d_a a$).

Proof. Let us start with the basic theta function

(5.9)
$$\theta_0(X) = \prod_{n \geq 0}(1 - q^nX)\prod_{n>0}(1 - q^nX^{-1}) \quad ,$$

(this is the theta function Θ_4 with the notations of Jacobi). Its divisor is the only point 1 (mod $q^{\mathbb{Z}}$) with multiplicity 1, representing a simple zero of this function at all points of $q^{\mathbb{Z}}$: $\operatorname{div}\theta_0 = (1)$.

One checks easily $\theta_o(q^{-1}X) = -X\theta_o(X)$ whence

$$d_q((1)) = 1 \ , \quad \phi_q((1)) = 1 \bmod q^{\mathbb{Z}} \ .$$

For any $a \in K^\times$, let us put $\theta_a(X) = \theta_o(a^{-1}X)$, so that $\operatorname{div}\theta_a = (a)$. Then $\theta_a(q^{-1}X) = (aq)^{-1}(-X)\theta_a(X)$, and so

$$d_q((a)) = 1 \ , \quad \phi_q((a)) = a \bmod q^{\mathbb{Z}} \ .$$

If $\underline{d} = \sum d_a(a) \in \mathrm{Div}_K(E_q)$ is any divisor on E_q (rational over K), the theta function $\prod \theta_a^{d_a}$ has divisor $\underline{d}$, hence one gets

$$d_q(\underline{d}) = \sum d_a \cdot d_q(\theta_a) = \sum d_a = \deg \underline{d} \ ,$$

$$\phi_q(\underline{d}) = \prod \phi_q(\theta_a)^{d_a} = \prod a^{d_a} \bmod q^{\mathbb{Z}} \ ,$$

as asserted.

(5.10) Corollary 1. A divisor $\underline{d} \in \mathrm{Div}_K(E_q)$ is principal (i.e. of the form $\underline{d} = \operatorname{div} f$ for some $f \in L_K^q$) precisely when $d_q(\underline{d}) = \deg \underline{d} = 0$ and $\phi_q(\underline{d}) = 1 \bmod q^{\mathbb{Z}}$.

(5.11) Corollary 2. If $a \in K^\times$ and b is not conjugate to a mod $q^{\mathbb{Z}}$ (over K), there exists a principal divisor $\underline{d}$ with multiplicities $d_a = 1$ and $d_b = 0$.

Proof. Consider first the case $a \in K^\times$ and take for $\underline{d}$ the divisor $\underline{d} = (a) - (au) - ((v) - (vu))$ with $u,v \in K^\times$ chosen so that all a, au, v, vu, b are distinct mod $q^{\mathbb{Z}}$ ($K^\times/q^{\mathbb{Z}}$ is infinite). This divisor is principal because it satisfies the two conditions of corollary 1. When on the contrary $a \notin K^\times$, let $d = [K(a) : K] \geqslant 2$ and $N = N_{K(a)/K}(a) \in K^\times$. Then $\underline{d}_a = \sum (a^\sigma)$ is rational over K and of degree d. Moreover $\phi_q(\underline{d}_a) = N$. If now $u,v \in K^\times$ are chosen such that $N = u^{d-1}v$ and u, v, b, a^σ are all distinct mod $q^{\mathbb{Z}}$, then

$$\underline{d} = \underline{d}_a - ((d-1)(u) + (v)) \in \mathrm{Div}_K(E_q)$$

satisfies the conditions of corollary 1, so is principal.

As usual, for $\underline{d} \in \mathrm{Div}_K(E_q)$, we define the K-subspace $L_K(\underline{d}) = L_K^q(\underline{d}) \subset L_K^q$ by $\operatorname{div}(f) \geqslant -\underline{d}$ (with the convention $\operatorname{div}(0) \geqslant -\underline{d}$

for all divisors $\underline{d}$). These are finite dimensional vector spaces over K and the fact that L_K^q is an elliptic function field will follow from the following more precise result.

(5.12) Theorem (Riemann-Roch). For every divisor $\underline{d} \in \mathrm{Div}_K(E_Q)$ of positive degree $d > 0$, $\dim_K L_K(\underline{d}) = d$.

Proof. If $d = 1$, a reduction which we have used repeatedly brings us back to the case $\underline{d} = (a)$ has only one multiplicity $d_a = 1 \neq 0$. Then corollary 1 shows that $L_K((a)) = K$ is of dimension one. Now, we can use induction on $d = \deg \underline{d} > 1$. Put $a = \phi_q(\underline{d})$ and select $b \in K^*$ $(d_b=0)$ distinct from 1 and a mod $q^{\mathbb{Z}}$. Define $\underline{d}' = \underline{d} - (b) \in \mathrm{Div}_K(E_q)$ of degree $d' = d - 1 > 0$. By induction hypothesis $L(\underline{d}')$ is of dimension d' (over K). As it is the kernel of the linear mapping

$$f \longmapsto f(b) \ , \ L(\underline{d}) \longrightarrow K \ ,$$

it is sufficient to show the surjectivity of this mapping. But the divisor $(a) + (d-1)(1) - \underline{d} = \mathrm{div}(f)$ is principal and has multiplicity 0 at b : $f(b) \neq 0$. This proves the theorem.

Note that if K' is a finite algebraic extension of K, then $H_{K'} = H_K \cdot K'$ and so $M_{K'} = M_K \cdot K'$ (if $f/g \in M_{K'}$ is a quotient of elements of $H_{K'}$ with $g \neq 0$, amplify this fraction by the conjugates of the denominator so as to be reduced to the case $g \in H_K$: then use $f \in H_{K'} = H_K \cdot K'$). Choosing a basis (e_i) of K' over K, write any $f \in L_{K'}^q \subset M_{K'}$ in the form $f = \sum f_i e_i$ with some $f_i \in M_K$. Because $f(q^{-1}X) = f(X)$ and the linear independence of the e_i (K' is linearly disjoint from M_K over K, e.g. because we may assume K'normal over K), we infer $f_i(q^{-1}X) = f_i(X)$, i.e. $f_i \in L_K^q$. This proves $L_{K'}^q = L_K^q \cdot K'$. Then the theorem above applied for K' instead of K shows that

$$\dim_{K'} L_{K'}^q(\underline{d}) = \deg \underline{d} \quad \text{if} \quad \deg \underline{d} > 0 \ .$$

This proves that L_K^q is of genus one over K ($L_K^q \cdot \bar{K} = \bigcup_{K'} L_{K'}^q$ is of genus one over $\bar{K}$). In particular, L_K^q is (isomorphic to) the field of

K-rational functions over a non-singular (absolutely irreducible) cubic defined over K. I contend that moreover E_q is the set of normalized valuations of $L_K^q \cdot \bar{K}$ trivial over $\bar{K}$:

(5.13) $$E_q \longrightarrow X(V) \longrightarrow V(\bar{K}) \quad ,$$
$$a \longmapsto \mathrm{ord}_a \longmapsto P_a = \text{center of } \mathrm{ord}_a \quad ,$$

where $\mathrm{ord}_a(f)$ is defined transcendentally with Schnirelmann's theorem as the multiplicity of the zero (or pole) of f at a. That these mappings are injective is obvious (V is non singular). The surjectivity of the first one (or the composite) is easily seen : if $P \in V(\bar{K})$ was not in the image, then take a finite extension $K' = K(P)$ of K such that $P \in V(K')$, and $f \in L_{K'}(2(P))$ not constant. This leads to a contradiction because Schnirelmann's theorem shows that $\mathrm{ord}_a f < 0$ for some $a \in E_q$. Moreover, because V(K) is the set of points fixed by $\mathrm{Gal}(\bar{K}/K)$, $X_K(V)$ is the set of valuations which are invariant under all $\sigma \in \mathrm{Gal}(\bar{K}/K)$ and in the isomorphism (5.13) we see that the inverse image of V(K) is precisely

(5.14) $$K^\times/q^{\mathbb{Z}} = E_q(K) \subset E_q = E_q(\bar{K}) = \bar{K}^\times/q^{\mathbb{Z}} .$$

From now on, we shall identify E_q with the corresponding non-singular cubic curve V and write for example $E_q(K')$ instead of $V(K')$ for any extension K' of K .

The main problem left over is that of determining the invariant of the elliptic function field L_K^q over K, as a function of q in the (punctured) unit disc $|q| < 1$ of $K^\times$. This will be done now by looking at the Weierstrass functions $\wp$ and $\wp'$ as functions of the p-adic variable q (instead of the usual variable z).

To find the p-adic analogue, let us come back to some classical computations regarding the $\wp$ function of Weierstrass, relative to the lattice $L = \mathbb{Z} + \tau\mathbb{Z}$. By definition

$$\wp(z:\tau) = z^{-2} + \sum_{(m,n)\neq(0,0)} \left\{ (z-m-n\tau)^{-2} - (m+n\tau)^{-2} \right\},$$

with a normal convergence on all bounded sets not containing lattice points. We sum this series by keeping first n fixed and using the classical formula $(\pi/\sin\pi z)^2 = \sum (z-m)^{-2}$ (summation extended over all rational integers $m \in \mathbb{Z}$). For $n \neq 0$ we get

$$\sum_m \left\{(z-m-n\tau)^{-2} - (m+n\tau)^{-2}\right\} = \sum_m (z-n\tau-m)^{-2} - \sum_m (n\tau+m)^{-2},$$

because both series converge, and the sum is then

$$(5.15) \qquad (\pi/\sin\pi(z-n\tau))^2 - (\pi/\sin\pi n\tau)^2 .$$

The terms corresponding to $n = 0$ give similarly

$$z^{-2} + \sum_{m\neq 0} \left\{(z-m)^{-2} - m^{-2}\right\} = \sum_m (z-m)^{-2} - 2\sum_1^{\infty} m^{-2} =$$

$$= (\pi/\sin\pi z)^2 - \pi^2/3 .$$

Gathering this expression with (5.15) for all $n \neq 0$, we get

$$(5.16) \qquad \wp(z:\tau) = \sum (\pi/\sin\pi(z+n\tau))^2 - \pi^2/3 - \sum_{n\neq 0} (\pi/\sin\pi n\tau)^2 .$$

Now we introduce the new variables

$$(5.17) \qquad X = \underline{e}(z) = e^{2\pi i z} \quad \text{and} \quad q = \underline{e}(\tau) = e^{2\pi i\tau} .$$

We see that

$$(\pi/\sin\pi(z+n\tau))^2 = (2\pi i)^2 q^n X/(1-q^n X)^2 ,$$

and thus

$$\wp(z:\tau) = (2\pi i)^2 \left[\sum_n q^n X/(1-q^n X)^2 + 1/12 - \sum_{n\neq 0} q^n/(1-q^n)^2\right]$$

$$= (2\pi i)^2 P(X) ,$$

with

$$(5.18) \qquad P(X) = P(X:q) = \sum q^n X/(1-q^n X)^2 + 1/12 - 2\sum_1^{\infty} q^n/(1-q^n)^2 .$$

We can still express the last term in this expression slightly differently. Using the binomial formula, we get indeed

$$(1 - q^n)^{-2} = 1 + 2q^n + 3q^{2n} + \dots \quad ,$$
$$q^n(1 - q^n)^{-2} = q^n + 2q^{2n} + 3q^{3n} + \dots \quad ,$$

so that

$$q^n/(1 - q^n)^2 = \sum_{n\geq 1}\sum_{m\geq 1} mq^{mn} = \sum_{m\geq 1} m \sum_{n\geq 1} q^{mn} =$$
$$= \sum_{m\geq 1} mq^m/(1 - q^m) = s_1(q) \ ,$$

where we use the more general notation

(5.19) $$s_k(q) = \sum_{m\geq 1} m^k q^m/(1 - q^m) = \sum_{N\geq 1} \sigma_k(N) q^N \qquad (k \in \mathbf{N}) \ ,$$

and $\sigma_k(N)$ denotes the sum of the kth powers of all divisors of N .

Thus we have

(5.20) $$P(X) = \sum q^n X/(1 - q^n X)^2 + 1/12 - 2s_1(q) \ .$$

To express the differential equation satisfied by this function P we note that

$$(2\pi i)^{-1} d/dz = X\, d/dX = D \qquad \text{(say)}$$

and so $\wp' = (2\pi i)^2 (2\pi i) DP = (2\pi i)^3 DP$

$$(2\pi i)^6 (DP)^2 = \wp'^2 = 4\wp^3 - g_2\wp - g_3 = 4(2\pi i)^6 P^3 - (2\pi i)^2 g_2 P - g_3$$

and finally

(5.21) $$(DP)^2 = 4P^3 - g_2' P - g_3' \qquad \text{with}$$
$$g_2' = (2\pi i)^{-4} g_2 \quad , \quad g_3' = (2\pi i)^{-6} g_3 \quad .$$

From (5.20) we also get

(5.22) $$DP(X) = DP(X:q) = \sum (q^n X + q^{2n} X^2)/(1 - q^n X)^3 \ ,$$

and from the classical expressions given in the first chapter (I.4.2-3)

(5.23) $$g_2' = (1 + 240 s_3(q))/12 \ , \ g_3' = -(1 - 504 s_5(q))/216 \ ,$$
$$j = (12)^3 g_2^3/\Delta = (12)^3 (g_2')^3/\Delta' = q^{-1} + \sum_{n\geq 0} c(n) q^n \ ,$$
$$\Delta' = (2\pi)^{-12}\Delta = q \prod_{n\geq 1} (1 - q^n)^{24}$$

(look at (I.4.5.b) for the last formula).

The advantage of these expressions is that the factors π have been cleaned out, that their coefficients (integers divided by some powers of 2 and 3 : $216 = 6^3$) have a universal meaning, and in particular make sense over p-adic fields. Returning now to the p-adic case with $0 \neq q \in K$, $|q| < 1$, we define two q-elliptic functions by (5.20) and (5.22) respectively noting that these series converge due to the fact that their general terms tend to 0. The first one has a pole of order 2 at 1 (mod $q^{\mathbb{Z}}$) whereas the second one has a pole of order 3 at the same point (and they are regular elsewhere). These functions must generate L_K^q over K (by the Riemann-Roch theorem). In fact all their coefficients are in $\mathbb{Q}(q) \subset K$. The complex identity (5.21) gives a formal identity in X and q with coefficients in $\mathbb{Z}[1/6] \subset \mathbb{Q} \subset K$ (5.23), hence will also be satisfied p-adically. This proves in particular that the absolute invariant $j = j(E_q)$ is given by the convergent series

(5.24) $$j(q) = 1/q + 744 + \sum_{n \geq 1} c(n) q^n \qquad (c(n) \in \mathbb{Z})$$

having the same coefficients as the classical one. Because the c(n) are integers, their p-adic absolute value is smaller or equal to 1, and $\sum_{n \geq 0} c(n) q^n$ will itself be a p-adic integer (the subring of integers in K, defined by $|x| \leq 1$, is compact, hence complete) and we infer from that, that $j(E_q)$ cannot be a p-adic integer :

(5.25) $$|j(E_q)| > 1 \quad .$$

In fact, for any $j \in K$ with $|j| > 1$, there exists a unique $q \in K$ with $|q| < 1$ and $j = j(q)$. This follows from the next lemma which uses essentially the ultrametric property of the absolute value of K.

(5.26) Lemma. Let $f(X) = \sum_{n \geq 0} c_n X^n$ be a formal series with coefficients $c_n \in K$ satisfying $|c_n| \leq 1$. Then the mapping $x \mapsto 1/x + f(x)$ defines a continuous bijection between the punctured unit disc $0 < |x| < 1$ and the exterior of that disc $|x| > 1$.

Proof. First we observe that the series converges for $|x| < 1$ by hypothesis of integrality of the coefficients c_n. Hence $x \mapsto x^{-1} + f(x)$ is well-defined and continuous in $0 < |x| < 1$. Let us check now the

<u>injectivity</u> of the mapping. If $x^{-1} + f(x) = y^{-1} + f(y)$ we derive

$$(y-x)(xy)^{-1} = x^{-1} - y^{-1} = f(y) - f(x) =$$
$$= \sum_{n\geq 1} c_n (y^n - x^n) = \sum_{n\geq 1} c_n (y-x)(y^{n-1} + \dots + x^{n-1}) ,$$

and hence because we assume $|x| < 1$, $|y| < 1$ all terms $y^{n-1} + \dots + x^{n-1}$ are in the unit disc and so is their sum :

$$|y-x|\ |xy|^{-1} \leqslant |y-x| \left|\sum \dots\right| \leqslant |y-x| .$$

This implies indeed $|y-x| = 0$, $y = x$. To show the <u>surjectivity</u>, we fix any $y \in K$ with $|y| > 1$ and we solve $y = 1/x + f(x)$ for x by iteration. This equation is equivalent to $y = (1 + xf(x))/x$ and also to $x = y^{-1}(1 + xf(x))$. Thus we define inductively a sequence $(x_i)_{i\geq 0}$ by $x_0 = 0$, $x_{i+1} = y^{-1}(1 + x_i f(x_i))$. By induction one checks $|x_i| < 1$ (and more precisely $|x_i| = |y^{-1}|$ for $i \geq 1$). Then

$$x_{i+1} - x_i = y^{-1}(x_i f(x_i) - x_{i-1} f(x_{i-1})) =$$
$$= y^{-1} \sum_{n\geq 0} c_n (x_i - x_{i-1})(x_i^n + \dots + x_{i-1}^n) ,$$

hence by iteration

$$|x_{i+1} - x_i| \leqslant |y^{-i}|\,|x_1 - x_0| = r^i |x_1| \qquad (r = |y^{-1}| < 1)$$

showing that (x_i) is a Cauchy sequence (we are in an ultrametric space !) . The limit is the required solution.

We know that two elliptic curves over K are isomorphic over $\bar{K}$ if and only if they have the same absolute invariant j. We may ask questions about K-isomorphism classes of elliptic curves. More precisely, we may ask : when are the two elliptic curves (g_i , $g_i' \in K$)

$$y^2 = 4x^3 - g_2 x - g_3 \ , \ y^2 = 4x^3 - g_2' x - g_3' \quad \text{with } j = j'$$

isomorphic over K (When do they give rise to K-isomorphic function fields?). The answer is given by (2.36). When the g_i's are not 0, a necessary and sufficient condition is that

$$g_2'/g_3' = t^2(g_2/g_3) \text{ with some } t \in K$$

(indeed this is equivalent to $g_2' = t^{-4} g_2$ and $g_3' = t^{-6} g_3$) .

For the curves defined by a Weierstrass equation with $g_2g_3 \neq 0$ we define thus the <u>relative invariant</u>

(5.27) $$\gamma = -\tfrac{1}{2}(g_2/g_3) \bmod K^{\times 2} .$$

This is a well-defined element of $K^{\times}/K^{\times 2}$, independent of the particular Weierstrass equation chosen with coefficients in K which (together with the absolute invariant j) characterizes completely the K-isomorphism class of the elliptic curve. The reason for the choice of the factor $-\tfrac{1}{2}$ will be apparent below. To compute the relative invariant of Tate's curve E_q , we need a lemma.

(5.28) <u>Lemma</u>. <u>Let</u> K <u>be a</u> p-<u>adic field</u>, $x \in K$ <u>with</u> $|x| < 1$. <u>Then</u> $1 + 4x$ <u>is a square of</u> K : $1 + 4x \in K^{\times 2}$.

<u>Proof</u>. We start with the formal series expansion

$$(1+4X)^{\frac{1}{2}} = 1 + \tfrac{1}{2}(4X) + \tfrac{1}{2}(-\tfrac{1}{2})(4X)^2/2! + \ldots$$

with coefficient of X^k (up to the sign) given by

$$1\cdot 3\cdot 5 \ldots (2k-3)2^k/k! = 2\cdot\ 1\cdot 3\cdot 5\ldots(2k-3)\cdot 2\cdot 4\cdot 6\ldots(2k-2)/(k!(k-1)!)$$

$$= 2\,\frac{(2k-2)!}{k!(k-1)!} = \frac{2}{k}\binom{2(k-1)}{k-1} .$$

But

$$\binom{2k}{k} = (2k-1)\frac{2}{k}\binom{2(k-1)}{k-1} \quad \text{is an integer ,}$$

and k is prime to 2k-1 (e.g. because 2k - (2k-1) = 1) so that k must divide $2\binom{2(k-1)}{k-1}$. This proves that the coefficients of $(1+4X)^{\frac{1}{2}}$ are all integers : $(1+4X)^{\frac{1}{2}} = \sum_{n\geq 0} a_nX^n$ ($a_n \in \mathbb{Z}$) which gives a convergent power series expression $\sum a_nx^n$ for the square root of $1+4x$.

Because both $1 + 240s_3$ and $1 - 504s_5$ are of the form $1+4x$ with $|x|<1$, these are squares of K and the relative invariant of Tate's curve E_q (as given by the canonical Weierstrass equation (5.21)), using (5.23), is

$$\gamma = +\tfrac{1}{2}\,\frac{216}{12}(1 + 240s_3)(1 - 504s_5)^{-1} = 9y^2 ,$$

and is a square. We have chosen the factor $-\tfrac{1}{2}$ precisely to trivialize the relative invariant of Tate's curve. We sum up our results in the main theorem of this section.

(5.29) Theorem (Tate). Let $q \in K$ (p-adic field of characteristic 0) be such that $0 < |q| < 1$, and E_q be Tate's elliptic curve $\bar{K}^{\times}/q^{\mathbb{Z}}$. Then

a) E_q is defined over K with set of K-rational points $E_q(K)$ isomorphic to $K^{\times}/q^{\mathbb{Z}}$.

b) The absolute invariant of E_q is given by the convergent series

$$j(E_q) = q^{-1} + 744 + \sum_{n \geq 1} c(n) q^n$$

(with the coefficients c(n) of the classical q-expansion of j), so that in particular $|j(E_q)| > 1$.

c) The relative invariant $\gamma = -\frac{1}{2} g_2/g_3$ of a Weierstrass equation of E_q over K is trivial : $\gamma \in K^{\times 2}$.

Conversely, for every $|j| > 1$ in K, there is a unique curve E_q with invariant j.

Observe that because $|j| > 1$, j is never 0 nor 12^3 for Tate's curves, and consequently neither g_2 nor g_3 vanishes. Then, our definition of the relative invariant is meaningful. For characteristics $p \neq 2,3$ this theorem is valid without modification (the only denominators appearing in (5.23) are powers of 2 and 3) because Schnirelmann's theorem is valid in all characteristics. In characteristic 2 (and 3), some other normalizations have to be adopted (as is explained in the book by P. Roquette on Analytic Theory of Elliptic Functions over Local Fields).

(5.30) Remark. The expression (5.20) for P(X) could essentially have been derived from the basic theta function θ_0 defined in (5.9) in the usual way (this is the method chosen by Roquette). First, the logarithmic derivative gives the analogue to the ξ function of Weierstrass. Taking the derivative once more gives substantially the P function :

$$-D(D\theta_o/\theta_o) = \sum_{\mathbb{Z}} q^n X/(1 - q^n X)^2 =$$

$$= P(X) - 1/12 + 2s_1(q) \quad .$$

The fact that we do not get exactly the P-function comes from the definition of θ_o without convergence factors (they are needed in the classical theory in the σ-function, but are superfluous in the p-adic case).

CHAPTER THREE

Division Points

Division points on a complex torus play a role analogous to roots of unity on the circle. In particular, there are plenty of them, and if the torus is given as cubic with rational coefficients, their coordinates are algebraic numbers, hence their algebraic interest. By taking suitable limits of groups of division points, some canonical p-adic spaces are attached to the curve, which, formally at least, are similar to the tangent space at the origin :

$V_\infty(E) = \mathrm{Lie}(E) = \mathrm{Hom}(\mathbb{R},E)$ is of $\mathrm{dimension}_{\mathbb{R}}$ two ,

$T_p(E) = \mathrm{Hom}(\mathbb{Q}_p/\mathbb{Z}_p, E)$ free $\mathbb{Z}_p$-module of rank two ,

$V_p(E) = T_p(E) \otimes_{\mathbb{Z}_p} \mathbb{Q}_p$ is of $\mathrm{dimension}_{\mathbb{Q}_p}$ two .

Apart from the basic definitions and properties, we have also indicated some applications. In particular, if L is a sublattice of an imaginary quadratic field, we have proved in two ways that its invariant j(L) is an algebraic integer. The first one is the classical analytical one, whereas the second one (Tate) uses division points through ℓ-adic representations of Tate's curves.

Prerequisites for this chapter are still quite limited. For the sake of simplicity we have only treated Hecke correspondences of prime level, so that we have to use the fact that a quadratic imaginary field always contains an element of prime norm.

1. Division points in characteristic zero

Let us start with some p-adic preliminaries. Every p-adic number $x \in \mathbb{Q}_p$ has a well-defined polar part $\langle x \rangle$ of the form q/p^n $(0 \leqslant q < p^n)$ defined by the property

$$x \in \langle x \rangle + \mathbb{Z}_p \quad (\text{equivalently } x \equiv \langle x \rangle \bmod \mathbb{Z}_p) \ .$$

If $x = \sum a_m p^m$ $(0 \leqslant a_m < p)$, then $\langle x \rangle$ is the finite sum $\sum_{m<0} a_m p^m$. It is obvious that $\langle x + x' \rangle = \langle x \rangle + \langle x' \rangle \bmod \mathbb{Z}$ and that the mapping $x \longmapsto \langle x \rangle$ is locally constant on $\mathbb{Q}_p$ (it is constant on the additive cosets of $\mathbb{Z}_p$). This proves that

$$x \longmapsto \underline{e}_p(x) = \underline{e}\langle x \rangle = \exp(2\pi i \langle x \rangle)$$

is a (non-trivial continuous) character of $\mathbb{Q}_p$. This character $\underline{e}_p$ is called <u>Tate's canonical character of</u> $\mathbb{Q}_p$. In particular, $\underline{e}_p$ identifies $\mathbb{Q}_p/\mathbb{Z}_p$ with the subgroup $\underline{e}(p^{-\infty}\mathbb{Z}) \subset \mathbb{C}^1 \subset \mathbb{C}^\times$ generated by the roots of 1 having as order a power p^n of the prime p.

(1.1) <u>Proposition 1</u>. <u>For</u> $x \in \mathbb{Q}_p$, <u>let</u> $\hat{x}$ <u>be the character of</u> $\mathbb{Q}_p$ <u>defined by</u> $y \mapsto \underline{e}_p(xy)$. <u>Then</u> $x \mapsto \hat{x}$ <u>defines an isomorphism of</u> $\mathbb{Q}_p$ <u>onto its</u> (<u>Pontryagin</u>) <u>topological dual</u>.

<u>Proof</u>. The homomorphism $x \mapsto \hat{x}$ is injective because $\hat{x} = 0$ means $\underline{e}_p(xy) = 1$ for all $y \in \mathbb{Q}_p$ or $xy \in \mathbb{Z}_p$ for all y, hence $x = 0$. Its image is dense in the character group by duality, because if $y \in \mathbb{Q}_p$ is orthogonal to all characters $\hat{x}$, it implies as above $y = 0$. Let us see that it is <u>bicontinuous</u>. The continuity of $x \mapsto \hat{x}$ is clear. Conversely, let us show that $\hat{x} \to 0$ implies $x \to 0$. Let $V_r(R)$ be the neighbourhood of 0 in the dual of $\mathbb{Q}_p$ defined by

$$|\hat{x}(y) - 1| = |\underline{e}_p(xy) - 1| < r \text{ for } |y|_p \leqslant R \ .$$

Fix $u \in \mathbb{Q}_p$ with $\underline{e}_p(u) \neq 1$ and choose $r \leqslant |\underline{e}_p(u) - 1|$ so that if $\hat{x} \in V_r(R)$ then $x^{-1}u = y$ cannot be in the ball of radius R (by the definition of

$V_r(R)$ just given) : $|x^{-1}u|_p > R$ and $|x|_p < |u|/R$. This shows that when $R \to \infty$ and $\hat{x} \in V_r(R)$, necessarily $x \to 0$. As consequence, the image of $\mathbb{Q}_p$ under $x \longrightarrow \hat{x}$ is a locally compact subspace of this dual and must be closed (because it is a subgroup). This proves the surjectivity of the mapping.

(1.2) Corollary. If we identify $\mathbb{Q}_p$ with its topological dual, $\mathbb{Z}_p$ is its own orthogonal, and the dual of this compact group is (isomorphic to) the discrete group $\mathbb{Q}_p/\mathbb{Z}_p$. In other words the exact sequence

$$0 \longrightarrow \mathbb{Z}_p \longrightarrow \mathbb{Q}_p \longrightarrow \mathbb{Q}_p/\mathbb{Z}_p \longrightarrow 0$$

is autodual, i.e. gives by duality the same sequence

$$0 \longleftarrow \mathbb{Q}_p/\mathbb{Z}_p \longleftarrow \mathbb{Q}_p \longleftarrow \mathbb{Z}_p \longleftarrow 0 \quad .$$

Proof. The assertion on the orthogonal of $\mathbb{Z}_p$ follows directly from the fact that the kernel of $\underline{e}_p$ is $\mathbb{Z}_p$. For the remaining assertion, just use the general fact that if H is a closed subgroup of the locally compact abelian group G, then the dual $\hat{H}$ of H is canonically isomorphic to the quotient $\hat{G}/H^{\perp}$ (of course in our case $G = \mathbb{Q}_p$, $H = \mathbb{Z}_p$ and this general result could be derived more directly !).

(1.3) Proposition 2. There is a canonical isomorphism $\mathbb{Q}/\mathbb{Z} \longrightarrow \bigoplus_p \mathbb{Q}_p/\mathbb{Z}_p$. In other words, the p-primary component of $\mathbb{Q}/\mathbb{Z}$ is (isomorphic to) $\mathbb{Q}_p/\mathbb{Z}_p$.

Proof. The composite homomorphism $\mathbb{Q} \longrightarrow \mathbb{Q}_p \longrightarrow \mathbb{Q}_p/\mathbb{Z}_p$ is trivial on $\mathbb{Z}$, hence a homomorphism $\mathbb{Q}/\mathbb{Z} \longrightarrow \mathbb{Q}_p/\mathbb{Z}_p$. The product of these gives a homomorphism $\mathbb{Q}/\mathbb{Z} \longrightarrow \prod_p \mathbb{Q}_p/\mathbb{Z}_p$ whose image is contained in the direct sum $\bigoplus_p \mathbb{Q}_p/\mathbb{Z}_p$ (the subgroup of families having only finitely non-zero components) because a rational number has only finitely many primes dividing its denominator. hence belongs to $\mathbb{Z}_p$ for nearly all primes p . The injectivity follows from the fact that for $x \in \mathbb{Q}$,

$$x \in \mathbb{Z}_p \text{ (when considered in } \mathbb{Q}_p) \text{ for all } p \Longrightarrow x \in \mathbb{Z} \ .$$

Finally the elements $q/p^n \in \mathbb{Q}/\mathbb{Z}$ have images integral at all $\ell \neq p$ hence generate the component $\mathbb{Q}_p/\mathbb{Z}_p$.

(1.4) Proposition 3. The automorphism group of $(\mathbb{Q}/\mathbb{Z})^n$ is canonically isomorphic to $\prod_p Gl_n(\mathbb{Z}_p)$.

Proof. Because automorphisms respect the decomposition in p-primary components, we are reduced by Prop.2 above, to determining the automorphism groups of $(\mathbb{Q}_p/\mathbb{Z}_p)^n$. But by transposition, the automorphism group of $(\mathbb{Q}_p/\mathbb{Z}_p)^n$ is isomorphic to the automorphism group of the dual $\mathbb{Z}_p^n$. But now $Aut(\mathbb{Z}_p^n) = Gl_n(\mathbb{Z}_p)$ because every automorphism of the group $\mathbb{Z}_p^n$ is $\mathbb{Z}$-linear, hence also $\mathbb{Z}_p$-linear by continuity. In fact, every matrix in Gl_n with p-adic integral entries determines a $\mathbb{Q}_p$-linear mapping $\mathbb{Q}_p^n \longrightarrow \mathbb{Q}_p^n$ leaving stable the subgroup $\mathbb{Z}_p^n$, hence an automorphism of the quotient. This gives explicitely the isomorphism of Prop.3 .

Now we go back to our business and let k be a field of characteristic 0 , $E = E_{\bar{k}}$ be an elliptic curve over k given in Weierstrass form (or simply a non-singular projective plane cubic over k with one selected rational point over k). We denote by t(E) the torsion subgroup of E, and for any integer $N \geqslant 1$, by $t_N(E)$ (or more simply by ${}_NE$) its subgroup of elements x having an order divisible by N : $N\cdot x = 0 \in E$. For example we have already seen that

$t_2(E) \approx (\mathbb{Z}/2\mathbb{Z})^2$ set of ramification points of $\wp : E \longrightarrow \mathbf{P}^1$,

$t_3(E) \approx (\mathbb{Z}/3\mathbb{Z})^2$ set of flexes of E if in Weierstrass normal form (cf. (I.1.26) and (II.2.9-10)) . We are going to prove a more general result.

(1.5) Proposition 4. For any integer $N \geqslant 1$, the subgroup $t_N(E) \subset E$ of division points satisfying $N\cdot x = 0$ is isomorphic to $(\mathbb{Z}/N\mathbb{Z})^2 = (N^{-1}\mathbb{Z}/\mathbb{Z})^2$. The full torsion subgroup is

$$t(E) = \bigcup_{N\geq 1} t_N(E) = \varinjlim t_N(E) = (\mathbb{Q}/\mathbb{Z})^2 = \bigoplus_p (\mathbb{Q}_p/\mathbb{Z}_p)^2 .$$

Proof. We may assume k is of finite type over the prime field $\mathbb{Q}$, hence we may choose an embedding $\bar{k} \longrightarrow \mathbb{C}$ and with this choice $E \subset E_{\mathbb{C}}$. By the transcendental theory, $E_{\mathbb{C}}$ is isomorphic to a complex torus $\mathbb{C}/L$ with a certain lattice $L \subset \mathbb{C}$ and in particular

$$t_N(E_{\mathbb{C}}) = N^{-1}L/L \quad ,$$

$$t(E_{\mathbb{C}}) = \mathbb{Q}\cdot L/L \quad .$$

It remains to see $t(E_{\mathbb{C}}) \subset E = E_{\bar{k}}$, or equivalently $t_N(E_{\mathbb{C}}) \subset E$ for all integers $N \geq 1$. But $N\cdot x = 0$ is the equation for an algebraic variety V over k (which might be reducible), and we have seen that $V_{\mathbb{C}} \cap E_{\mathbb{C}}$ is finite. For any automorphism $\sigma \in \mathrm{Aut}(\mathbb{C}/k)$, V and E are invariant under σ (because defined over k), hence

$$(V_{\mathbb{C}} \cap E_{\mathbb{C}})^{\sigma} = V_{\mathbb{C}} \cap E_{\mathbb{C}} \quad .$$

This proves that these intersection points can only have finitely many conjugates under the automorphism group in question. These coordinates must thus be algebraic :

$$t_N(E_{\mathbb{C}}) = V_{\mathbb{C}} \cap E_{\mathbb{C}} \subset E_{\bar{k}} \quad ,$$

and this completes the proof. Observe that the proof shows that the field generated over k by the coordinates of the points of $t_N(E)$ is a Galois extension of k, and the same is true if we consider only the first (or second) coordinates of the points of $t_N(E)$ (in such a statement, we consider the affine coordinates of the points not at infinity if the curve E is given in Weierstrass normal form; we could also consider the projective coordinates, normalizing one of them to 1).

If E and E' are two elliptic curves as above, and if h is a homomorphism $E \longrightarrow E'$, then it induces a homomorphism, still denoted by h, between the corresponding torsion subgroups $h : t(E) \longrightarrow t(E')$.

For example, this gives a representation of the endomorphism ring of E : $\mathrm{End}(E) \longrightarrow \mathrm{End}(t(E)) = \prod_p M_2(\mathbb{Z}_p)$ (cf. Prop.3 above). In practice, it is more conceptual to let the endomorphisms of $t(E)$ act on a vector space over $\mathbb{Q}_p$, and for that purpose, to introduce the action on the topological dual of $t(E)$ (or any isomorphic space) and to extend the scalars to $\mathbb{Q}_p$. There is a canonical construction due to Tate which we explain briefly.

Let, for a moment, G be any additive p-divisible group : $pG = G$. We define the Tate module

(1.6) $$T_p(G) = \varprojlim t_{p^n}(G) \quad .$$

By definition, $x = (x_n)_{n\geqslant 0} \in T_p(G)$ means that $px_{n+1} = x_n$ and $p^n x_n = 0$ for all integers $n \geqslant 0$. This is a $\mathbb{Z}_p$-module because $t_{p^n}(G)$ is a module over $\mathbb{Z}/p^n\mathbb{Z} = \mathbb{Z}_p/p^n\mathbb{Z}_p$ and $\varprojlim \mathbb{Z}_p/p^n\mathbb{Z}_p = \mathbb{Z}_p$. Because there is a canonical isomorphism

$$\mathrm{Hom}(p^{-n}\mathbb{Z}/\mathbb{Z}, G) = t_{p^n}(G) \quad ,$$

another possible definition for Tate's module would have been

(1.7) $$T_p(G) = \varprojlim \mathrm{Hom}(p^{-n}\mathbb{Z}/\mathbb{Z}, G) = $$
$$= \mathrm{Hom}(\varinjlim p^{-n}\mathbb{Z}/\mathbb{Z}, G) = \mathrm{Hom}(\mathbb{Q}_p/\mathbb{Z}_p, G) \quad .$$

We could still replace G by its subgroup $G_p = t_{p^\infty}(G)$ of elements having a power of p as order (p-primary component of G) in this last formula. Now we consider the projective sequence of homomorphisms of multiplication by p

$$\cdots \xrightarrow{p} G_p \xrightarrow{p} G_p \xrightarrow{p} G_p \quad .$$

We put

(1.8) $$V_p(G) = V_p(G_p) = \varprojlim (G_p \xrightarrow{p} G_p) \quad ,$$

and call it the extended Tate module of G (or G_p , relative to the prime p). By definition $T_p(G) \subset V_p(G)$ is the submodule consisting of sequences $x = (x_n)$ with $x_0 = 0$, and the projection $x \longmapsto x_0$ onto the

0th component gives an exact sequence

$$0 \longrightarrow T_p(G) \longrightarrow V_p(G) \longrightarrow G_p \longrightarrow 0 .$$

Because each $x \in V_p(G)$ is such that $p^m x$ has 0th component $p^m x_0 = 0$ for m sufficiently large, we see that

$$V_p(G) = \bigcup_{m \geq 0} p^{-m} T_p(G) \quad ,$$

and because $T_p(G)$ has no torsion (multiplication by p is the shift operator), we have

$$V_p(G) = T_p(G) \underset{\mathbb{Z}_p}{\otimes} \mathbb{Q}_p \quad . \tag{1.9}$$

In particular, this extended module is a vector space over $\mathbb{Q}_p$.

Let us take in particular for G a torus $(\mathbb{R}/\mathbb{Z})^n \cong (\mathbb{C}^1)^n$. Then

$$\begin{aligned} T_p(G) &\cong \mathrm{Hom}(\mathbb{Q}_p/\mathbb{Z}_p, \mathbb{R}/\mathbb{Z})^n \cong \mathrm{Hom}(\mathbb{Q}_p/\mathbb{Z}_p, \mathbb{C}^1)^n = \\ &\cong (\text{top.dual of } \mathbb{Q}_p/\mathbb{Z}_p)^n \cong \mathbb{Z}_p^n \end{aligned} \tag{1.10}$$

(and this is also isomorphic to $\mathrm{Hom}((\mathbb{Q}_p/\mathbb{Z}_p)^n, \mathbb{C}^1) \cong$ top.dual of $t_{p^\infty}(G)$).

This is a free $\mathbb{Z}_p$-module of rank n and consequently $V_p(G)$ is a vector space of dimension n over $\mathbb{Q}_p$ in this case. These spaces can be looked at as p-adic analogues of the tangent space (elements of order p are "closer to the origin" than elements of order p^2, in the algebraic sense). For $p = \infty$, $\mathbb{Q}_p = \mathbb{Q}_\infty = \mathbb{R}$ and we could put

$$V_\infty(G) = \mathrm{Hom}(\mathbb{R}, G) = \mathrm{Lie}(G) = \mathfrak{g} \quad .$$

To come back to our case, we let $G = E = E_{\bar{k}}$ be the group of $\bar{k}$-rational points on our elliptic curve. There are two canonical representations attached to the space $V_p(E)$ (2-dimensional over $\mathbb{Q}_p$). The first one is a representation of the ring of endomorphisms of E

$$\mathrm{End}(E) \longrightarrow \mathrm{End}(V_p(E)) \quad , \tag{1.11}$$

and the second one is a representation of the Galois group of the algebraic (separable) closure of k over k

$$\mathrm{Gal}(\bar{k}/k) \longrightarrow \mathrm{Aut}(V_p(E)) \quad . \tag{1.12}$$

These two representations are the main reason for introducing the p-adic modules $T_p(E)$ and $V_p(E)$. In particular, the Galois module $V_p(E)$ is isomorphic to the vector dual (over $\mathbb{Q}_p$) of the etale cohomology group $H^1_p(E)$ defined by Artin-Grothendieck. It could thus be called first p-adic homology group of E. In a particular case (transcendental invariant j), the image of $Gal(\bar{k}/k)$ in $Aut(V_p(E))$ will be determined explicitely in the next section.

From a somewhat different point of view, let us add a few general considerations on rational division points over an elliptic curve E defined over a number field k, i.e. those with coordinates in k. More precisely, we can show that the k-rational torsion subgroup $t(E_k)$ of E is finite (this would follow from the Mordell-Weil theorem asserting that E_k is finitely generated, but we prove this corollary directly by a local method). Let $\mathfrak{p}$ ($\neq 0$) be a prime ideal in the ring of integers $\mathfrak{o}_k$ of k, which we assume prime to (2) for the sake of simplicity. We denote by $K = k_{\mathfrak{p}}$ its $\mathfrak{p}$-adic completed field, by R the ring of integers of K and by P the maximal ideal of the local ring R. The announced result will follow from the following local uniformization theorem for $\mathfrak{p}$-adic elliptic curves.

(1.13) Proposition. Let E be an elliptic curve defined over the p-adic field K, say by a Weierstrass equation $y^2 = 4x^3 + ax + b$ with integral coefficients $a,b \in R$. Then there is an open subgroup U of E_K, isomorphic both algebraically and topologically to the (additive) group of integers R of K.

Proof. The neighbourhood in question on the elliptic curve E will be defined by x large, or 1/x small, and because infinity is a ramification point of index two for x we shall take $t = 1/x^{\frac{1}{2}}$ as local uniformizing variable. To be able to do that, we have to check that 1/x is indeed a square in our p-adic field K. But

$$1/x = (2x/y)^2(1 + a/4x^2 + b/4x^3) \quad ,$$

and this is a square in K as soon as x is big, say $x^{-1} \in \mathfrak{p}$, because the formal series for $(1 + X)^{\frac{1}{2}}$ is convergent in $\mathfrak{p}$ (the denominators of its coefficients have only powers of 2 in their denominators, hence are in R : we use the fact that $\mathfrak{p}$ is prime to (2) here, otherwise the convergence radius of that series would be smaller). Thus let us put $x = t^{-2}$, whence $y^2 = 4t^{-6} + at^{-2} + b = t^{-6}(4 + at^4 + bt^6)$ and

$$1/y = \pm t^3(4 + at^4 + bt^6)^{-\frac{1}{2}} = \pm(t^3/2 + \text{higher order terms})$$

is given by a power series with coefficients in R. More precisely, the coefficients of this series are in $\mathbb{Z}[\frac{1}{2},a,b] \subset R$ (polynomials in a and b with rational coefficients having only powers of two in their denominators). Let us now define by a formal term by term integration

$$z(t) = \int \frac{dx}{y} = \int (t^3/2 + \ldots)(-2t^{-3})dt = -t + \ldots$$

Now if we call a_n the $n^{\underline{th}}$ coefficient of that series, $na_n \in \mathbb{Z}[\frac{1}{2},a,b]$, because the integration has introduced the denominator n. This series has same convergence radius as that giving 1/y because

$$\operatorname{ord} \frac{x^n}{n} = n \operatorname{ord} x - \operatorname{ord} n \longrightarrow \infty \qquad \text{if } \operatorname{ord} x > 0 \quad ,$$

and so $|x^n/n| = |x|^n/n \longrightarrow 0$ if $|x| < 1$. If we solve now for t, we will find $t = -z + b_2z^2 + b_3z^3 + \ldots$ and the equations giving the b_i recursively as polynomials in the b_j $(j < i)$ and a_k $(k \leqslant i)$ show that $n!b_n \in \mathbb{Z}[\frac{1}{2},a,b]$ (compare with the expansions of the logarithm and the exponential functions). Thus the series giving $t = t(z)$ will converge in the same disc as the exponential series

$$t(z) = \sum_{n\geqslant 1} \frac{p_n(a,b)}{n!} z^n \quad , \quad p_n(a,b) \in \mathbb{Z}[\tfrac{1}{2},a,b] \; .$$

But the exponential series has a non-zero convergence radius as follows from the well-known formula

$$\operatorname{ord}_p(n!) = \frac{n - S(n)}{p - 1} \quad \text{if } S(n) = \sum n_i \text{ for } n = \sum n_i p^i \; .$$

I claim now that $z \longmapsto (x(t(z)),y(t(z))) = ((t(z)^{-2},-2t(z)^{-3}, \ldots)$ is a group isomorphism in the domain of convergence of these series. But this assertion amounts to a lot of identities between the coefficients $p_n(a,b)$ of $t= t(z)$. To check these identities, we choose an embedding $a \longmapsto A, b \longmapsto B$ of $\mathbb{Z}[\frac{1}{2},a,b]$ into the complex field $\mathbb{C}$. then we just observe that they result from the classical (complex) theory of elliptic curves, where they are true formally (i.e. when A and B are indeterminates) because we could choose A and B transcendental, independent. This proves the polynomial identities and the isomorphism in the domain of convergence of the series $t = t(z)$, where all p-adic series have a meaning.

(1.14) Corollary 1. Let $j \in K$ be fixed. Then there is a constant M_j such that the order $\mathrm{Card}\, t(E_K) \leqslant M_j$ for every elliptic curve E defined over K of invariant j.

Proof. The projective space $\mathbb{P}^2(K)$ is compact because it can be covered by the three compact charts, images of the compact sets

$$\Omega_j = \{(x_i)_{i=0,1,2} : x_i \in R \ , \ x_j = 1\} .$$

This implies that the closed subset E_K is also compact. On the other hand E_K has an open subgroup U isomorphic to R, hence without torsion. Consequently

$$\mathrm{Card}\, t(E_K) \leqslant \mathrm{Card}\, E_K/U \quad \text{is finite} \quad .$$

Now the K-isomorphism classes of elliptic curves of invariant j are parametrized by the finite sets

$K^\times/(K^\times)^2$ if $j \neq 0,1728$ (relative invariant (II.5.27)),

$K^\times/(K^\times)^4$ if $j = 1728$,

$K^\times/(K^\times)^6$ if $j = 0$.

This gives the uniformity of the bounds for the orders of $t(E_K)$ for fixed invariant j.

(1.15) Corollary 2. Let E be an elliptic curve defined over the number field k. Then the order of the rational torsion subgroup $t(E_k)$ is finite.

Proof. Observe that with the above notations $t(E_k) \subset t(E_K)$!

(1.16) Remark. It has been conjectured for some time that the finite number of rational torsion points Card $t(E_k)$ on elliptic curves defined over k is bounded by a constant M_k depending only on the number field k and not the elliptic curve E over k (with $M_k \longrightarrow \infty$ for increasing $k \subset \bar{k} = \bar{\mathbb{Q}}$). Manin first proved a weak form of that conjecture, showing that for any prime number p, the $p^{\underline{th}}$ component of that order is bounded (uniformly in E defined over k). Recently, the conjecture in its strong form has been proved by Demjanenko.

2. An ℓ-adic representation of a Galois group

Let E be an elliptic curve defined over a field k of characteristic 0. If $j = j(E)$ is the absolute invariant of E, necessarily $\mathbb{Q}(j) \subset k$. We suppose that $k \subset \mathbb{C}$ is embedded in the complex field. For any integer $N \geqslant 1$, we denote by $t_N(E) = {}_NE$ the subgroup of points t of E having an order dividing N : $N \cdot t = 0$. This is a free $\mathbb{Z}/N\mathbb{Z}$ - module of rank 2. There is a canonical $\mathbb{Z}$-bilinear form over this module, with values in the group $\underline{e}(N^{-1}\mathbb{Z}/\mathbb{Z})$ of $N^{\underline{th}}$ roots of 1 in $\mathbb{C}$ which can be defined as follows. For $t \in {}_NE$, the divisor $N((t) - (0))$ over E is of degree 0 and satisfies Abel's condition, hence is principal. Take a rational function over E, $f_t \in \mathbb{C}(E)$ with this divisor (f_t is determined up to a multiplicative constant by this condition) $\mathrm{div}(f_t) = N((t) - (0))$. Select $t' \in E$ with $N \cdot t' = t$ (noting that two such points t' and t'' must differ by a point in ${}_NE$: $t'' = t' + u$ with $u \in {}_NE$). The divisor

$$\underline{d}_t = \sum_{u \in {}_NE} \left[(t'+u) - (u)\right] \in \mathrm{Div}(E)$$

depends only on t and not on the choice of t' with $N \cdot t' = t$, has degree 0, and satisfies Abel's condition :

$$\sum_{u \in {}_NE} (t'+u-u) = \sum_{u \in {}_NE} t' = N^2 t' = N \cdot t = 0 .$$

This proves that $\underline{d}_t$ is also principal, and we can find a rational function $F_t \in \mathbb{C}(E)$ on E with divisor $\underline{d}_t$ (and this condition determines F_t up to a multiplicative constant). Then

$$\mathrm{div}(F_t^N) = N\underline{d}_t = \sum_{{}_NE} N((t'+u) - (u))$$

is the divisor of the rational function $f_t \circ N : v \longmapsto f_t(N \cdot v)$ so that there exists a constant $c \in \mathbb{C}^\times$ (depending on t, the choices of f_t and F_t) with $F_t^N(v) = c f_t(N \cdot v)$, and replacing v by $v + s$ (for $s \in {}_NE$), we get $F_t^N(v+s) = F_t^N(v)$. Hence there exists a well-defined $N^{\underline{th}}$ root

$e_N(t,s)$ with $F_t(v+s) = e_N(t,s)F_t(v)$ (for all $v \in E$). This mapping e_N has the following properties.

(2.1) <u>Proposition</u>. <u>The mapping</u> $e_N : {}_NE \times {}_NE \longrightarrow {}_N\mu = \underline{e}(N^{-1}\mathbb{Z}/\mathbb{Z})$ <u>is</u> $\mathbb{Z}$-<u>bilinear and satisfies</u>

a) e_N <u>is antisymmetrical</u> : $e_N(s,t) = e_N(t,s)^{-1}$,

b) e_N <u>is non-degenerate</u> : $e_N(t,s) = 1$ <u>for all</u> $s \in {}_NE$ <u>implies</u> $t = 0 \in E$,

c) <u>for any automorphism</u> σ <u>of</u> $\mathbb{C}$ <u>trivial on</u> k (<u>or on any field of definition of</u> E) $e_N(t,s)^\sigma = e_N(t^\sigma,s^\sigma)$.

<u>Proof</u>. For brevity we write $e = e_N$ in this proof, hoping that no confusion will arise ! By definition, it is clear that $e(t,s+s') = e(t,s)e(t,s')$. Let us prove that also $e(t+t',s) = e(t,s)e(t',s)$. For that, put $t'' = t+t'$ and take a rational function $F \in \mathbb{C}(E)$ with $\operatorname{div}(F) = (t) + (t') - (t'') - (0)$. Then

$$\operatorname{div}(F^N) = N(t) + N(t') - N(t'') - N(0) = \operatorname{div}(f_t f_{t'}/f_{t''})$$

with certain choices of functions f_t, $f_{t'}$ and $f_{t''}$ corresponding to the points t, t', t''. Thus

$$(F_t^N F_{t'}^N/F_{t''}^N)(v) = (cf_t c'f_{t'}/c''f_{t''})(N\cdot v) = C\cdot F^N(N\cdot v)$$

and hence

$$(F_t F_{t'}/F_{t''})(v) = C'F(N\cdot v)$$

is invariant under the substitution $v \longmapsto v+s$ ($s \in {}_NE$). This proves $e(t,s)e(t',s)/e(t'',s) = 1$ hence the $\mathbb{Z}$-bilinearity. We turn to the proof of a). For $n = 1,\dots,N$ define the translate $f_{t,n}$ of f_t by $f_{t,n}(v) = f_t(v-nt)$, and compute the divisor of the product $\prod f_{t,n}$

$$\operatorname{div}(\prod_{n=1}^{N} f_{t,n}) = N\sum_{n=1}^{N}\left[(t+nt) - (nt)\right] = 0 .$$

Hence $\prod_{n=1}^{N} f_{t,n} = C$ is constant and

$$\prod f_{t,n}(v) = \prod f_{t,n}(N\cdot v) = \prod f_t(Nv-nt) = \prod f_t(N(v-nt'))$$

is equal by definition to $\prod c_n F_t^N(v-nt')$ and must be constant.

Extracting the Nth root, we see that $\prod_{n=1}^{N} F_t(v-nt')$ must be constant and replacing v by $v+t'$ we get

$$\prod_{n=1}^{N} F_t(v-nt') = \prod_{n=1}^{N} F_t(v+t'-nt') =$$
$$= \prod_{n=0}^{N-1} F_t(v-nt') \quad ,$$

and after simplification by the common factors, $F_t(v) = F_t(v-Nt') = F_t(v-t)$ is invariant by translation of t : $e(t,t) = 1$ (for any $t \in {}_N E$). From there, replacing t by $t+s$ and using the bilinearity of the symbol $e = e_N$, we derive

$$1 = e(t+s,t+s) = e(t,s)e(s,t) \ .$$

For the proof of b), note that if $e(t,s) = 1$ for all $s \in {}_N E$, we have $F_t(v+s) = F_t(v)$ for all these s and so $F_t(v) = \phi(Nv)$ implying $F_t^N = (\phi \circ N)^N = f_t \circ N$ so that $\operatorname{div}(\phi \circ N) = N^{-1}\operatorname{div}(f_t \circ N)$ and

$$\operatorname{div}(\phi) = \operatorname{div}(f_t)/N = (t) - (0) \ .$$

Abel's condition gives $t = 0$ so that $e = e_N$ is non-degenerate.

Finally, for c), we observe that σ acts on E and on the set X of normalized valuations of $\mathbb{C}(E)$ (trivial on $\mathbb{C}$) according to

$$(\operatorname{ord}_P)^\sigma (f^\sigma) = \operatorname{ord}_P(f) \qquad (\operatorname{ord}_P \in X \text{ or } P \in E).$$

In particular, if x and y are coordinate functions on E regular at P (and defined over k)

$$(\operatorname{ord}_P)^\sigma (x - x(P^\sigma)) = (\operatorname{ord}_P)^\sigma (x - x(P))^\sigma = \operatorname{ord}_P(x - x(P)) > 0 \ ,$$

and similarly $(\operatorname{ord}_P)^\sigma (y - y(P^\sigma)) > 0$, which proves that $(\operatorname{ord}_P)^\sigma$ is centered at P^σ : $(\operatorname{ord}_P)^\sigma = \operatorname{ord}_{P^\sigma}$, $\operatorname{ord}_{P^\sigma}(f^\sigma) = \operatorname{ord}_P(f)$. This shows that if $f \in \mathbb{C}(E)$, $\operatorname{div}(f^\sigma) = (\operatorname{div} f)^\sigma$ (with the natural action of σ on divisors). If $t,s \in {}_N E$ and f_t , F_t are chosen as before, we see that we can choose $f_{t^\sigma} = (f_t)^\sigma$, $F_{t^\sigma} = (F_t)^\sigma$ so that

$$e(t^\sigma,s^\sigma)F_{t^\sigma}(v^\sigma) = F_{t^\sigma}(v^\sigma + s^\sigma) = F_t^\sigma((v+s)^\sigma) = F_t(v+s)^\sigma =$$
$$= e(t,s)^\sigma F_t(v)^\sigma = e(t,s)^\sigma F_{t^\sigma}(v^\sigma) \ .$$

This proves $e(t,s)^\sigma = e(t^\sigma,s^\sigma)$ as asserted in c).

To be able to go to the inverse limit in the symbols e_N (with $N = \ell^n$) we have to give a connection between two of them.

(2.2) Proposition. Let M,N be two (strictly) positive integers, $t \in t_N(E) \subset t_{MN}(E)$, $s \in t_{MN}(E)$. Then one has

$$e_{MN}(t,s) = e_N(t,Ms) \qquad (Ms \in t_N(E)).$$

Proof. We simplify the notations for the proof, denoting by $e = e_N$, $e' = e_{MN}$ (putting primes ' to all notions relative to MN). We have used the notations

$$\operatorname{div} f_t = N(t) - N(0) \quad \text{and} \quad (F_t)^N = cf_t \circ N \quad .$$

From there we deduce

$$(F_t \circ M)^{MN} = (F_t^N \circ M)^M = c^M(f_t \circ NM)^M = c^M(f_t^M \circ MN) \quad .$$

Hence we may choose

$$f'_t = f_t^M \quad \text{(implying } \operatorname{div} f'_t = MN(t) - MN(0) \text{ as it should be)},$$

$$F'_t = F_t \circ M \quad , \ c' = c^M \quad ,$$

so that $(F'_t)^{MN} = c' f'_t\, MN$. Now by definition of e'

$$F'_t(v+s) = e'(t,s)F'_t(v) \quad ,$$

giving $F_t(Mv + Ms) = e'(t,s)F_t(Mv)$. But by definition of $e(t,s)$

$$F_t(w + Ms) = e(t,Ms)F_t(w) \quad ,$$

hence the result.

Taking now $t \in t_{MN}(E)$, $s \in t_{MN}(E)$, using the bilinearity, we get immediately

$$e_{MN}(t,s)^M = e_{MN}(Mt,s) = e_N(Mt,Ms).$$

Now we take for M and N powers of the prime ℓ . Remembering that an element of $T_\ell(E)$ can be identified with a sequence $(t_n)_{n\geq 0}$ with $t_n \in t_{\ell^n}(E)$ and $\ell \cdot t_n = t_{n-1}$, we define a bilinear symbol

$$e(t,s) = \langle t,s \rangle = (e_{\ell^n}(t_n,s_n))_{n\geq 0}$$

(for $t = (t_n)$ and $s = (s_n)$) and we consider the result in the projective limit $\varprojlim_{\ell^n} \mu \cong \varprojlim \mathbb{Z}/\ell^n\mathbb{Z} = \mathbb{Z}_\ell$. Thus $\langle \cdot , \cdot \rangle$ defines a non-degenerate antisymmetrical $\mathbb{Z}_\ell$-bilinear pairing

(2.3) $$T_\ell(E) \times T_\ell(E) \longrightarrow \mathbb{Z}_\ell \ , \ (t,s) \longmapsto \langle t,s \rangle \ .$$

We shall need the following easy lemmas.

(2.4) Lemma 1. Let D be a domain in $\mathbb{C}$ (not empty open connected subset) and (f_i) an at most denumerable family of meromorphic functions in D. If k is a demumerable subfield of $\mathbb{C}$, then there is a point $z_0 \in D$ such that the mapping $f \mapsto f(z_0)$ (is well defined and) gives an embedding of the field $k(f_i)_i$ generated over k by the f_i into the complex field.

Proof. The field of meromorphic functions $k(f_i)_i$ is denumerable, and each meromorphic function has at most denumerably many poles. Since D is open (hence not denumerable), there exists a point z_0 which is pole for no function $f \in k(f_i)$. The mapping $f \mapsto f(z_0)$ is then well-defined, not zero, hence gives an isomorphism $k(f_i) \longrightarrow \mathbb{C}$ (into). A mapping of the form $f \mapsto f(z_0)$ (defined on a subring of a certain function field) is called a specialization. When moreover the homomorphism $f \mapsto f(z_0)$ is injective (hence gives an isomorphism into $\mathbb{C}$) the specialization (or the point z_0 itself) is said to be generic. With this terminology, lemma 1 asserts that every denumerable function field of meromorphic functions on a domain $D \subset \mathbb{C}$ has a generic specialization. (Lemma 1 is also true for a denumerable function field of meromorphic functions on a domain D in $\mathbb{C}^n$, using Baire's theorem - D is locally compact - because the pole sets of meromorphic functions have no interior points.)

(2.5) Lemma 2. Let $\alpha \in Gl_2(\mathbb{Z}/N\mathbb{Z})$ be a transformation with the property $\alpha(t) = \pm t$ for every $t \in (\mathbb{Z}/N\mathbb{Z})^2$. Then $\alpha = \pm 1$.

Proof. Call t_1, t_2 the canonical basis of $(\mathbb{Z}/N\mathbb{Z})^2$ and define the signs $\varepsilon_i = \pm 1$ by $\alpha(t_i) = \varepsilon_i t_i$. Then $\alpha(t_1 + t_2) = \varepsilon_1 t_1 + \varepsilon_2 t_2$ must be $\pm(t_1 + t_2)$ by hypothesis. Hence $\varepsilon_1 = \varepsilon_2 = \varepsilon$ and $\alpha = \varepsilon \cdot 1$.

(2.6) Lemma 3. Reduction modulo N (a positive integer) defines a surjective homomorphism $Sl_2(\mathbb{Z}) \longrightarrow Sl_2(\mathbb{Z}/N\mathbb{Z})$. In other words, the following sequence is exact :

$$(1) \rightarrow \Gamma_N \rightarrow \Gamma = Sl_2(\mathbb{Z}) \rightarrow Sl_2(\mathbb{Z}/N\mathbb{Z}) \rightarrow (1)$$

(We denote by Γ_N the principal congruence subgroup of Γ of level N. By definition, it is a normal subgroup of Γ with finite index.)

Proof. We have to show that every matrix $\begin{pmatrix} a & b \\ c & d \end{pmatrix}$ with integral coefficients and determinant ad - bc $\equiv$ 1 mod N is congruent mod N (termwise) to a matrix with determinant 1. The elementary divisors theorem shows that there exists $\gamma, \delta \in Sl_2(\mathbb{Z})$ with $\gamma \begin{pmatrix} a & b \\ c & d \end{pmatrix} \delta = \begin{pmatrix} m & 0 \\ 0 & n \end{pmatrix}$ in diagonal form. Hence we are reduced to proving the theorem for diagonal matrices only. But if mn $\equiv$ 1 mod N we have the congruence

$$\begin{pmatrix} m & 0 \\ 0 & n \end{pmatrix} \equiv \begin{pmatrix} m & mn-1 \\ 1-mn & n(2-mn) \end{pmatrix} \mod N .$$

The matrix in the right-hand side is easily verified to have determinant 1. (In this lemma, Sl_2 could be replaced by Sl_n, the proof being made by induction on n with a similar method.)

Now we introduce Weber's functions. They are associated to division points on elliptic curves in the following way. Let $\mathrm{Im}(\tau) > 0$ and consider the complex torus $\mathbb{C}/L_\tau$. Its division points of order dividing N are the images of the points $a\tau/N + b/N$ with integers a and b. Introducing the line vector $i = (a/N, b/N) \in (N^{-1}\mathbb{Z}/\mathbb{Z})^2$ we can denote them by $i\binom{\tau}{1}$ (product of a line vector by a column vector, resulting in a scalar). Or, if we prefer to introduce the integral line vector $i' = Ni = (a,b) \in (\mathbb{Z}/N\mathbb{Z})^2$, we can also write these division points in the form $i'\binom{\tau/N}{1/N}$. Then Weber's functions are defined by

$$f_i(\tau) = \frac{g_2 g_3}{\Delta}(\tau)\, \wp(i\tbinom{\tau}{1} : L_\tau) \quad \text{if } 0 \neq i \in (N^{-1}\mathbb{Z}/\mathbb{Z})^2$$

(we discard an integral multiplicative constant which would play no role here). We also put $f_o = j$ (the modular function). In more algebraic terms, these Weber functions are normalized first projection of division points :

$$\begin{array}{ccl} \varphi : \mathbb{C}/L_\tau & \longrightarrow & E \subset \mathbb{P}^2(\mathbb{C}) \text{ defined by } y^2 = 4x^3 - g_2(\tau)x - g_3(\tau) \\ & \overset{\wp}{\searrow} & \downarrow x \\ & & \mathbb{P}^1(\mathbb{C}) \end{array}$$

is a commutative diagram, and if we put $\xi = \frac{g_2 g_3}{\Delta}(\tau)x$ for the normalized first projection of E, then

(2.7) $$f_i(\tau) = \frac{g_2 g_3}{\Delta}(\tau)\, \wp(i\binom{\tau}{1} : L_\tau) = \frac{g_2 g_3}{\Delta}(\tau) x(i'\binom{t_1}{t_2}) = \xi(i'\binom{t_1}{t_2}),$$

where

$$t_1 = \varphi(\tau/N) \ , \ t_2 = \varphi(1/N) \in E \ .$$

We observe that the rational function x on E being defined over $k = \mathbb{Q}(g_2(\tau), g_3(\tau))$ (field of definition of E) : $x \in k(E)$, the normalized first projection ξ is also rational over k : $\xi \in k(E)$. Had we taken any other Weierstrassian model E' of $\mathbb{C}/L_\tau$ with other coefficients $g_2' = \mu^4 g_2(\tau)$, $g_3' = \mu^6 g_3(\tau)$, the isomorphism

$$\psi : E \xrightarrow{\sim} E' \ , \ (x_o, y_o) \longmapsto (\mu^2 x_o, \mu^3 y_o)$$

shows that the normalized first projection ξ' of E' would take the same values as ξ on corresponding points :

(2.8) $$\xi'(t') = \xi(t) \text{ if } \psi(t) = t' \quad (\text{or } t = \varphi(z),\ t' = \varphi'(z)).$$

In particular, we can choose for E' a model defined over the smallest field possible : $k' = \mathbb{Q}(j(\tau))$.

(2.9) <u>Proposition</u>. <u>The field</u> $\mathbb{Q}(f_i : i \in (N^{-1}\mathbb{Z}/\mathbb{Z})^2)$ <u>of rational functions over the upper half-plane is a Galois extension of the field</u> $\mathbb{Q}(f_o) = \mathbb{Q}(j)$.

<u>Proof</u>. By lemma 1, there is a point τ in the upper half-plane, giving a generic specialization $f \longmapsto f(\tau)$ on this field, and consequently we have to prove the corresponding assertion for the extension

$$\mathbb{Q}(f_i(\tau))_i \ / \ \mathbb{Q}(j(\tau)) \ .$$

Choose then a model E', over the field $\mathbb{Q}(j(\tau))$, of the complex torus $\mathbb{C}/L_\tau$ *). By (2.7) and (2.8), noting that $i'\binom{t_1}{t_2}$ describes the set of points $t' \in E'$ annihilated by N, we get

(2.10) $$\mathbb{Q}(f_i(\tau))_i = \mathbb{Q}(j(\tau), \xi'(t') : 0 \neq t' \in t_N(E')).$$

This is a Galois extension of $\mathbb{Q}(j(\tau))$ because ξ' is rational over this field : any automorphism σ of the complex field $\mathbb{C}$, trivial over $j(\tau)$ will transform E' in itself, and $t_N(E')$ in itself with

$$\xi'(t'^\sigma) = \xi'(t')^\sigma \quad \text{(as well as } x'(t'^\sigma) = x'(t')^\sigma\text{)}.$$

This proves the proposition.

We shall determine later the Galois group of the extension $\mathbb{Q}(f_i)_i/\mathbb{Q}(f_0)$. But knowing that this extension is algebraic, we conclude that any τ (in the upper half-plane) such that $j(\tau)$ is transcendental is generic : the subring of $\mathbb{Q}(f_i)_i$ over which $f \longmapsto f(\tau)$ is well-defined (finite) must contain $\mathbb{Q}(j)$ and the f_i hence coïncide with $\mathbb{Q}(j)[f_i]_i = \mathbb{Q}(f_i)_i$.

(2.11) Corollary. Every τ (in the upper half-plane) such that $j(\tau)$ is transcendental, gives a generic specialization $f \longmapsto f(\tau)$ of the field $\mathbb{Q}(f_i)_i$ into the complex field $\mathbb{C}$.

It is easy to construct certain automorphisms of the field $\mathbb{Q}(f_i)_i$ (over $\mathbb{Q}(j)$). If $\gamma \in Sl_2(\mathbb{Z})$, $f \longmapsto f\circ\gamma$ gives one. Indeed, write $\gamma = \begin{pmatrix} a & b \\ c & d \end{pmatrix}$ and observe that the lattice $(c\tau + d)L_{\gamma(\tau)}$ is generated by the two vectors $a\tau + b$ and $c\tau + d$ hence is equal to L_τ. Then

(2.12)
$$\begin{aligned} f_i(\gamma(\tau)) &= \frac{g_2g_3}{\Delta}(\gamma(\tau))\, \wp(i\binom{\gamma(\tau)}{1} : L_{\gamma(\tau)}) = \\ &= \frac{g_2g_3}{\Delta}(\tau)\, \wp(i\binom{a\tau+b}{c\tau+d} : (c\tau+d)L_{\gamma(\tau)}) = f_{i\gamma}(\tau). \end{aligned}$$

Thus composition with the fractional linear transformation associated to γ permutes the indices (for $i \neq 0$ and obviously leaves $f_0 = j$ fixed) through right multiplication of the line index i by γ.

*) Observe that $j(\tau)$ must be transcendental, hence $g_2g_3(\tau) \neq 0$.

Because the $\wp$-function of Weierstrass (or the first projection of E) is even, we also see that $\gamma \in \Gamma_N$ will produce the trivial automorphism of $\mathbb{Q}(f_i)_i$, or equivalently the permutation $i \longmapsto \pm i$ of the indices, exactly when $\gamma \in \Gamma_N(\pm 1)$ (cf. lemma 2 above). Hence we get an <u>injective</u> homomorphism

(2.13) $$\Gamma/\Gamma_N(\pm 1) \cong Sl_2(\mathbb{Z}/N\mathbb{Z})/(\pm 1) \rightarrow Gal(\mathbb{Q}(f_i)_i/\mathbb{Q}(f_0)) \qquad (f_0 = j)$$
$$\pm\gamma \longmapsto (f_i \longmapsto f_i \circ \gamma = f_{i\gamma})$$

(cf.lemma 3).

Let E be an elliptic curve of invariant $j_0 \in \mathbb{C}$ defined over the minimal field $k = \mathbb{Q}(j_0)$. We look at the following Galois extensions of k :

$$\left.\begin{array}{l} K = k(t : t \in t_N(E)) \\ \quad | \\ \left.\begin{array}{l} K' = k(x(t) : t \in t_N(E)) \\ \quad | \\ \quad k \end{array}\right\} G' \end{array}\right\} G \quad .$$

Take a basis t_1, t_2 of the $\mathbb{Z}/N\mathbb{Z}$-module $t_N(E)$ and define for every automorphism $\sigma \in Gal(K/k)$ the matrix $\rho(\sigma)$ by

(2.14) $$\binom{t_1}{t_2}^{\sigma} = \binom{t_1^{\sigma}}{t_2^{\sigma}} = \rho(\sigma)\binom{t_1}{t_2} \quad .$$

Then ρ is an injective homomorphism

(2.15) $$\rho : G \longrightarrow Gl_2(\mathbb{Z}/N\mathbb{Z}) \quad .$$

If $\sigma \in Gal(K/K')$, then $x(t^{\sigma}) = x(t)$ implies $t^{\sigma} = \pm t$ for all $t \in t_N(E)$ hence $\rho(\sigma) = \pm 1$ by lemma 2. Indeed $t \longmapsto -t$ defines an automorphism of K over K' and this automorphism is not trivial if $N > 2$ which we shall suppose from now on. Then (2.15) gives also an injective homomorphism

(2.16) $$\rho' : G' \longrightarrow Gl_2(\mathbb{Z}/N\mathbb{Z})/(\pm 1) \;^{*)} \quad .$$

It is easy to see that ρ is surjective precisely when ρ' is : if ρ' is surjective, then either $\begin{pmatrix} 0 & -1 \\ 1 & 0 \end{pmatrix}$ or $\begin{pmatrix} 0 & 1 \\ -1 & 0 \end{pmatrix}$ is in the image of

*) If N = 2, K = K', G = G' and $(\pm 1) = 1$ so that (2.16) is also trivially true.

ρ , so that by taking the square, we find that -1 is in the image of ρ and so ρ is surjective. <u>Let us show that</u> K' <u>contains a primitive</u> $N^{\underline{th}}$ <u>root of</u> 1. Let $\zeta_N = e_N(t_1,t_2)$. Because the bilinear symbol e_N is non-degenerate, this is a primitive $N^{\underline{th}}$ root of 1 (if $\zeta_N^M = 1$, we see that $e_N(Mt_1,t_2) = 1$ hence $e_N(Mt_1,t) = 1$ for every $t \in t_N(E)$, and this implies $Mt_1 = 0$, i.e. M multiple of N). On the other hand, if $\sigma \in \mathrm{Aut}(\mathbb{C}/k)$, then

$$e_N(t_1,t_2)^\sigma = e_N(t_1^\sigma,t_2^\sigma) = e_N(t_1,t_2)^{\det\rho(\sigma)} ,$$

with the matrix $\rho(\sigma)$ of the restriction of σ to K. If σ is trivial on K', then we have just seen that $\rho(\sigma) = \pm 1$ so that $\det \rho(\sigma) = 1$ and $\zeta_N^\sigma = \zeta_N$. By Galois theory, this proves $\zeta_N \in K'$. But more precisely, the action of an automorphism $\sigma \in G$ (or $\sigma \in G'$) on ζ_N is given explicitely in the representation ρ (or ρ') by

(2.17) $$\zeta_N^\sigma = \zeta_N^{\det \rho(\sigma)} = \zeta_N^{\det \rho'(\sigma)} .$$

Because every automorphism of $k(\zeta_N) \subset K'$ over k extends to an automorphism of K' over k, we conclude that the image of ρ' (and ρ) contains elements of all determinants in $(\mathbb{Z}/N\mathbb{Z})^\times$. We have not used fully the assumption on the invariant j_0 so far. <u>But when</u> j_0 <u>is transcendent</u>, <u>we shall show that</u> ρ <u>and</u> ρ' <u>are surjective</u> (hence <u>isomorphisms</u>). Obviously, it only remains to show that in this case, the image of ρ' contains $\mathrm{Sl}_2(\mathbb{Z}/N\mathbb{Z})/(\pm 1)$. But the function j takes all complex values (once in the fundamental domain for the modular group), so that there exists a τ in the upper half-plane with $j(\tau) = j_0$. Because j_0 is transcendent, the specialization $f \longmapsto f(\tau)$ is generic for the field $\mathbb{Q}(f_i)_i$ (corollary (2.11)). We denote by $\sigma(\gamma)$ the automorphism of the specialized field corresponding to the automorphism of $\mathbb{Q}(f_i)$ over $\mathbb{Q}(j)$ given by (2.13), for $\gamma \in \mathrm{Sl}_2(\mathbb{Z}/N\mathbb{Z})/(\pm 1)$. Let E' be an algebraic model of $\mathbb{C}/L_\tau$ defined over $j(\tau) = j_0$. Then E' is isomorphic to E, and we can even take E' = E. By (2.7), the field $\mathbb{Q}(f_i(\tau))_i$ is the same as the field $\mathbb{Q}(j_0, x(t): 0 \neq t \in t_N(E)) =$ $= k(x(t): 0 \neq t \in t_N(E)) = K'$. Let us determine the representative matrix for the automorphism $\sigma(\gamma)$ of this field. We have for $i \neq 0$

$$f_i(\tau)^{\sigma(\gamma)} = f_i \circ \gamma(\tau) = f_{i\gamma}(\tau) = \xi(i'\gamma\binom{t_1}{t_2}) \quad ,$$

if the basis t_1, t_2 is chosen suitably (as in (2.7)). On the other hand,

$$f_i(\tau)^{\sigma(\gamma)} = \xi(i'\binom{t_1}{t_2})^{\sigma(\gamma)} = \xi(i'\begin{pmatrix} t_1^{\sigma(\gamma)} \\ t_2^{\sigma(\gamma)} \end{pmatrix}) = \xi(i'\rho(\sigma(\gamma))\binom{t_1}{t_2}) \ .$$

Comparison gives

$$i'\gamma = \pm i'\rho\sigma(\gamma) \quad \text{for all } i' \in (\mathbb{Z}/N\mathbb{Z})^2 \ .$$

This implies $\gamma = \rho \circ \sigma(\gamma) \in Sl_2(\mathbb{Z}/N\mathbb{Z})/(\pm 1)$ and shows that the image of ρ (or ρ') contains every element of determinant 1.

To sum up, we have proved

(2.18) <u>Theorem</u>. <u>Let</u> E <u>be an elliptic curve of transcendental invariant</u> $j_0 \in \mathbb{C}$ <u>defined over the minimal field</u> $k = \mathbb{Q}(j_0)$. <u>For every positive integer</u> $N \geqslant 1$, <u>the Galois representation</u>

$$\rho = \rho(N) : Gal(\bar{k}/k) \longrightarrow Aut(t_N(E))$$

<u>is surjective</u>. <u>If</u> ℓ <u>is any prime number</u>, <u>the</u> ℓ-<u>adic representation</u>

$$\rho_\ell : Gal(\bar{k}/k) \longrightarrow Aut(T_\ell(E)) \cong Gl_2(\mathbb{Z}_\ell)$$

<u>is surjective</u>.

Observe that the second assertion is deduced from the first one by taking $N = \ell^n$ and letting $n \to \infty$. The projective limit of these surjective homomorphisms is still surjective, because the image of $Gal(\bar{k}/k)$ must both be dense and compact (ρ_ℓ is continuous by definition).

Put $I = (\mathbb{Q}/\mathbb{Z})^2$ (line vectors) and consider the field (of rational functions on the upper half-plane) generated by all f_i ($i \in I$, remembering that $f_0 = j$ is the modular function) : $K_I = \mathbb{Q}(f_i)_I$. This field is a union of Galois extensions of $\mathbb{Q}(j) = \mathbb{Q}(f_0)$ hence is a Galois extension itself and the Galois group of $K_I/\mathbb{Q}(j)$ is the projective limit of the Galois groups of the subextensions $K'_N = \mathbb{Q}(f_i)_{t_N(I)}$, hence projective limit of the groups $Gl_2(\mathbb{Z}/N\mathbb{Z})/(\pm 1)$. Hence we have proved

(2.19) $$\mathrm{Gal}(K_I/\mathbb{Q}(j)) = \prod_p Gl_2(\mathbb{Z}_p) \;/\; (\pm 1) \;.$$

If we identify (by (1.3))

$$\mathbb{Q}/\mathbb{Z} = \bigoplus_p \mathbb{Q}_p/\mathbb{Z}_p \quad ,$$

we can say that the action of an element $\gamma \in \prod_p Gl_2(\mathbb{Z}_p)$ (mod ± 1) is given on the generators f_i through right multiplication of the index by the matrix :

$$f_i^{\sigma(\gamma)} = f_{i\gamma} \quad .$$

This follows from the fact that (2.16) is an isomorphism (onto) for every positive integer N, because if we choose a generic specialization $f \longmapsto f(\tau)$ for the field K_I (any τ with $j(\tau)$ transcendental will do), the action of a matrix on the index set I (at right) is transformed in the action on division points on an algebraic model defined over $\mathbb{Q}(j)$ of the complex torus $\mathbb{C}/L_\tau$. <u>The field K_I contains all roots of unity</u> (identified with constant functions on the upper half-plane). This follows by generic specialization from what has been seen on page III.18 . Write $\mathbb{Q}_{ab} = \mathbb{Q}(\underline{e}(1/N))_N$ for the maximal cyclotomic extension of $\mathbb{Q}$. Then we also have

(2.20) $$\mathrm{Gal}(\mathbb{Q}(f_i)_I/\mathbb{Q}_{ab}(j)) = \prod_p Sl_2(\mathbb{Z}_p) \;/\; (\pm 1) \;,$$

and by (2.12)

(2.21) $$\mathrm{Gal}(\mathbb{C}(f_i)_I/\mathbb{C}(j)) = \prod_p Sl_2(\mathbb{Z}_p) \;/\; (\pm 1) \quad .$$

The field $\mathbb{C}(j)$ is the field of all modular function (for Γ) by (I.3.11) and it is also easy to see that the field

$$\mathbb{C}(f_i)_{t_N(I)} = \mathbb{C}(f_i)_{i \in (N^{-1}\mathbb{Z}/\mathbb{Z})^2}$$

is the field of all automorphic functions of level N (with respect to the principal congruence subgroup Γ_N of Γ).

Let us write $\hat{\mathbb{Z}} = \prod_p \mathbb{Z}_p$ (ring of supernatural integers!) and make a picture of the embedding of these fields.

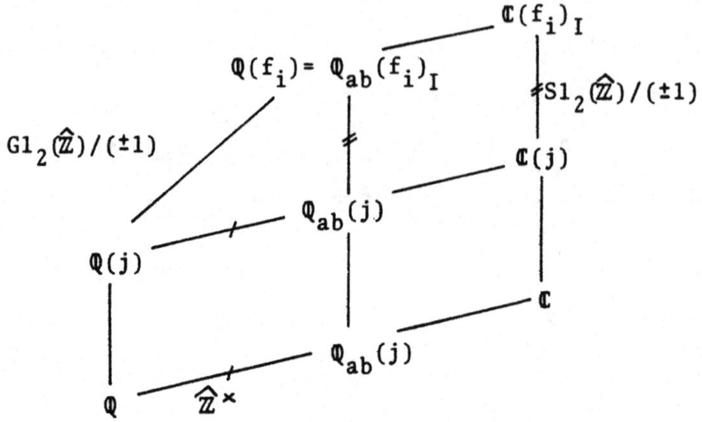

3. Integrality of singular invariants

The main goal of this section is to prove that if τ is a quadratic imaginary number (with positive imaginary part), then the value $j(\tau)$ of the modular invariant $j = 12^3 g_2^3/\Delta$ is an <u>algebraic integer</u>. We shall give two proofs of that fact : the classical proof which is transcendental and has some interesting features in itself, and a proof based on Tate's p-adic elliptic curves and their division points*) (which justifies the introduction of this section in this chapter !).

(3.1) <u>Definition. Let</u> $L \subset \mathbb{C}$ <u>be a lattice. We define its ring of endomorphisms by</u>

$$\mathrm{End}(L) = \{a \in \mathbb{C} : aL \subset L\} \supset \mathbb{Z} \quad ,$$

<u>and we say that</u> L <u>is singular when</u> $\mathrm{End}(L) \neq \mathbb{Z}$.

If we also define $L_{\mathbb{Q}} = \mathbb{Q}\cdot L \cong L \otimes_{\mathbb{Z}} \mathbb{Q}$ by extension of the scalars from $\mathbb{Z}$ to $\mathbb{Q}$ in the abelian group L , we also define the ring

$$\mathrm{End}(L_{\mathbb{Q}}) = \{a \in \mathbb{C} : aL_{\mathbb{Q}} \subset L_{\mathbb{Q}}\} \supset \mathbb{Q} \quad .$$

In fact, if $0 \neq a \in \mathrm{End}(L_{\mathbb{Q}})$, then $aL_{\mathbb{Q}} = L_{\mathbb{Q}}$, so that $a^{-1} \in \mathrm{End}(L_{\mathbb{Q}})$, and this ring $\mathrm{End}(L_{\mathbb{Q}})$ is a <u>field</u> (of characteristic 0). By definition it is clear that $\mathrm{End}(L_{\mathbb{Q}}) \supset \mathrm{End}(L)$ and that $\mathrm{End}(L) \cap \mathbb{R} = \mathbb{Z}$. This shows that L is singular precisely when $\mathrm{End}(L_{\mathbb{Q}}) \neq \mathbb{Q}$ is a strict extension of the rational field. The following lemmas are all immediate consequences of the definitions.

(3.2) <u>Lemma 1. For a normalized lattice</u> $L = L_\tau$ <u>to be a singular lattice, it is necessary and sufficient that</u> τ <u>be imaginary quadratic.</u>

<u>Proof</u>. If L_τ is singular, take $a \in \mathrm{End}(L)$ not an integer (hence $a \notin \mathbb{R}$). Then $aL_\tau \subset L_\tau$ implies $a = m + n\tau \in L_\tau$ (with some integers m, $n \neq 0$) and

---

*) communicated to me by J.-P. Serre.

also $a\tau \in L_\tau$ hence $(m + n\tau)\tau = p\tau + q$ which gives a quadratic equation satisfied by τ (we suppose $\mathrm{Im}(\tau) > 0$ hence τ is imaginary quadratic). Conversely, if τ is imaginary quadratic, $\tau^2 \in \mathbb{Q}L_\tau$ shows that $\tau(\mathbb{Q}L_\tau) \subset \mathbb{Q}L_\tau$ (equality in fact), so that $\mathrm{End}(\mathbb{Q}L_\tau) \neq \mathbb{Q}$.

(3.3) Lemma 2. L is a singular lattice whenever $\mathrm{End}(L_\mathbb{Q}) = K$ is an imaginary quadratic field, and $K = \mathbb{Q}(\tau)$ if $L = L_\tau$.

Proof. We may suppose $L = L_\tau$ and then we have seen that if L_τ is singular $\tau \in \mathrm{End}(\mathbb{Q}L_\tau)$, hence $K = \mathbb{Q}(\tau) \subset \mathrm{End}(\mathbb{Q}L_\tau)$. But $\mathbb{Q}L_\tau = \mathbb{Q}(\tau)$ proves that more precisely $\mathrm{End}(\mathbb{Q}L_\tau) = K$.

(3.4) Lemma 3. L is a singular lattice whenever End(L) is an order $\mathfrak{a}$ in the ring of integers of the quadratic field $K = \mathrm{End}(L_\mathbb{Q})$ (then L is homothetic to an ideal of this order).

Proof. This is obvious because the ring of integers is the maximal order of K (remember that an order is a free abelian group of maximal rank 2 which is also a ring with 1). Then L is isomorphic to a normal lattice L_τ contained in K, and a multiple of this lattice will be contained in the order End(L). Thus L is homothetic to an ideal of the order End(L) having precisely the order End(L) as ring of stabilizers (not an ideal of the ring of integers itself in general).

As to the structure of the orders in an imaginary quadratic field, we have the following lemma.

(3.5) Lemma 4. Let K be a quadratic field. The suborders $\mathfrak{a}$ of the ring of integers $\mathfrak{o}_K$ of K are the rings $\mathfrak{o}_f = \mathbb{Z} + f\mathfrak{o}_K$ with a positive integer f given by $f = [\mathfrak{o}_K : \mathfrak{a}]$ (finite because $\mathfrak{a}$ has rank 2).

Proof. Let $1, \omega$ be a basis of the free abelian group ($\mathbb{Z}$-module) $\mathfrak{o}_K$ and let $a + f\omega$ be an element of $\mathfrak{a}$ with minimal positive component of ω. Because $1 \in \mathfrak{a}$, $f\omega \in \mathfrak{a}$ and we shall have precisely $\mathfrak{a} = \mathbb{Z} + f\mathbb{Z}\omega = \mathbb{Z} + f\mathfrak{o}_K = \mathfrak{o}_f$. The index f of $\mathfrak{a}$ in $\mathfrak{o}_K$ is called the conductor of the order $\mathfrak{a}$ (it characterizes completely $\mathfrak{a}$ in our case).

We have $\mathfrak{O}_1 = \mathfrak{O}_K$ and an elementary computation shows that the discriminant D_f of $\mathfrak{O}_f$ is given by $D_f = f^2 D$ (where $D = D_1$ is the discriminant of $\mathfrak{O}_K$ or of $K/\mathbb{Q}$) .

Recalling that we say that two lattices are isomorphic if they are homothetic, we get the following result.

(3.6) Proposition. The classes of isomorphic singular lattices with fixed endomorphism ring $End(L) = \mathfrak{O}_f$ (order of conductor f in a given imaginary quadratic field K) are in one to one correspondence with the ideal classes of $\mathfrak{O}_f$ (with respect to the equivalence given by multiplication by principal ideals of $\mathfrak{O}_f$) having endomorphism ring (or ring of stabilizers) given exactly by $\mathfrak{O}_f$.

In particular, if $f = 1$ and h is the class number of K (order of the ideal class group of $\mathfrak{O}_K$), there are precisely h isomorphism classes of lattices $L \subset \mathbb{C}$ with $End(L) = \mathfrak{O}_K$. If more generally h_f denotes the number of ideal classes of $\mathfrak{O}_f$ having precisely $\mathfrak{O}_f$ as endomorphism ring, there will be h_f isomorphism classes of lattices L with $End(L) = \mathfrak{O}_f$. This ideal class group could be identified with I_f/P_f , the ideal class group (mod f) of fractional ideals of $\mathfrak{O}_K$ prime to f with respect to the subgroup of principal ideals $(\alpha) = \alpha\mathfrak{O}_f$ with $\alpha \equiv 1 \bmod f$. This is the classical ideal class group mod f of classfield theory, and in particular $h_f < \infty$ (for all $f \geqslant 1$). However, we do not need this (in particular, we shall not even need the finiteness of $h = h_1$).

(3.7) Theorem 1. If L is a singular lattice, $j = j(L)$ is an algebraic number.

Proof. This is the only point where we need elliptic curves for the main theorem of the section (in the transcendental method). Let $E = E_{\mathbb{C}} \simeq \mathbb{C}/L$ be the elliptic curve given by the equation

$$y^2 = 4x^3 - g_2 x - g_3 \quad \text{with } g_i = g_i(L) .$$

Then $\mathrm{End}(L) \cong \mathrm{End}(E)$ (ring of holomorphic endomorphisms of E keeping the origin - point at infinity - fixed : cf.(I.3.1) which shows that they are automatically homomorphisms) and any endomorphism $h : E \longrightarrow E$ is ipso facto algebraic, i.e. given by rational expressions (if p_i denotes the $i^{\underline{th}}$ projection $E \longrightarrow \mathbb{P}^1$, $i = 1,2$, coming from the embedding $E \subset \mathbb{P}^2$, then each $p_i \circ h : E \longrightarrow \mathbb{P}^1$ is an elliptic function on E, hence a rational function in $x = p_1$ ($= \wp$) and $y = p_2$ ($= \wp'$)). If now $\sigma \in \mathrm{Aut}(\mathbb{C})$ is any field automorphism and E^σ is the elliptic curve of equation

$$y^2 = 4x^3 - g_2^\sigma x - g_3^\sigma \quad ,$$

the mapping $h \longmapsto h^\sigma$ gives an isomorphism $\mathrm{End}(E) \longrightarrow \mathrm{End}(E^\sigma)$ (note that h^σ is again rational hence analytic, and thus a homomorphism by (I.3.1) because it leaves the point at infinity $(0,1,0) \in E^\sigma$ fixed, as did h for $(0,1,0) \in E$). This implies $E^\sigma \cong \mathbb{C}/L^\sigma$ with a lattice L^σ satisfying $\mathrm{End}(L^\sigma) = \mathrm{End}(L)$, hence isomorphic to a sublattice of $K = \mathrm{End}(L'_{\mathbb{Q}})$. Since there are only denumerably many such lattices, this implies that the set of isomorphism classes of E^σ ($\sigma \in \mathrm{Aut}(\mathbb{C})$) is at most denumerable. Because $j^\sigma = j(E^\sigma)$, this shows that the set $\{j^\sigma\}$ ($\sigma \in \mathrm{Aut}(\mathbb{C})$) of all conjugates of j is at most denumerable in $\mathbb{C}$. This proves that j is algebraic (otherwise it would have as many conjugates as the cardinal of a transcendental basis of $\mathbb{C}$ over $\mathbb{Q}$ which has the power of the continuum !). A posteriori, we see that j can have only finitely many conjugates. If one wants to use the fact that $h_f < \infty$, one deduces that the degree $[\mathbb{Q}(j):\mathbb{Q}]$ of j over $\mathbb{Q}$ is smaller or equal to h_f (f : conductor of End(E)) :

$$[\mathbb{Q}(j(\mathcal{O}_f)):\mathbb{Q}] \leqslant h_f \quad .$$

(3.8) Definition. A lattice L' is called isogenous to a lattice L when there exists $a \in \mathbb{C}^\times$ with $aL' \subset L$.

Isogeny is an equivalence relation (if $aL' \subset L$, then a multiple nL of L will be contained in aL', hence $(n/a)L \subset L'$). If L' is isogenous to L, then by definition $\mathrm{End}(L'_{\mathbb{Q}}) = \mathrm{End}(L_{\mathbb{Q}})$ because $L'_{\mathbb{Q}}$ is homothetic to $L_{\mathbb{Q}}$. Recall (I.3.4) that $L_{\tau'} = L_\tau$ when $\tau' = \tau + n = \eta(\tau)$ with some matrix $\eta \in \begin{pmatrix} 1 & \mathbb{Z} \\ 0 & 1 \end{pmatrix}$ and that $L_{\tau'} \cong L_\tau$ when $\tau' = \alpha(\tau)$ with some unimodular matrix $\alpha \in SL_2(\mathbb{Z})$ (= $GL_2(\mathbb{Z})_+$: group of inversible integral matrices with positive determinant).

(3.9) Lemma 5. Two lattices L_τ and $L_{\tau'}$ are isogenous exactly when $\tau' = \alpha(\tau)$ with some matrix $\alpha \in GL_2(\mathbb{Q})_+$ (rational matrix with positive determinant).

Proof. Take a complex number $t \neq 0$ such that $tL_{\tau'} \subset L_\tau$. Then

$$t\tau' = a\tau + b \text{ and } t = c\tau + d \text{ with } \alpha = \begin{pmatrix} a & b \\ c & d \end{pmatrix} \in M_2(\mathbb{Z}) .$$

Hence $\tau' = \alpha(\tau)$. Because τ and τ' have both positive imaginary parts $\det(\alpha) > 0$ (as in the proof of (I.3.4) we have more precisely $\mathrm{Im}(\tau') = \mathrm{Im}(\tau)\,|c\tau + d|^{-2}\det(\alpha)$). Conversely, if $\tau' = \alpha(\tau)$ with α as in the assertion, we can multiply it by an integer (without changing τ') and suppose $\alpha \in M_2(\mathbb{Z})$ to start with. If $\alpha = \begin{pmatrix} a & b \\ c & d \end{pmatrix}$, then $(c\tau + d)L_{\tau'} \subset L_\tau$.

As a consequence, the set of lattices with fixed $\mathrm{End}(L_{\mathbb{Q}}) = K$ (quadratic imaginary field), is a full equivalence class of isogenic lattices (L_τ is in this class whenever $K = \mathbb{Q}(\tau)$, i.e. $\tau \in K$, and two distinct $\tau, \tau' \in K$ with positive imaginary parts are linked by $\tau' = \begin{pmatrix} a & b \\ 0 & 1 \end{pmatrix}(\tau)$ with $a > 0, b \in \mathbb{Q}$).

(3.10) Theorem 2. If L' is isogenous to L and $j = j(L)$, $j' = j(L')$, then j' is integral over $\mathbb{Z}[j]$. (Note that we do not suppose L to be singular.)

The proof will be carried out in five steps.

First step : reduction to $\alpha = \begin{pmatrix} p & 0 \\ 0 & 1 \end{pmatrix}$ (p a prime number).

We can suppose $L = L_\tau$, $L' = L_{\alpha(\tau)}$ and by the elementary divisor theorem, there are unimodular matrices $\gamma, \delta \in SL_2(\mathbb{Z})$, and integers m,n , m divisible by n, so that $\alpha = \gamma \begin{pmatrix} m & 0 \\ 0 & n \end{pmatrix} \delta$. Using the invariance $j = j\circ\gamma = j\circ\delta$ of the modular function j, we can say that

$j\circ\alpha$ integral over $\mathbb{Z}[j]$ is equivalent to

$j\circ\gamma\circ\begin{pmatrix} m & 0 \\ 0 & n \end{pmatrix}\circ\delta = j\circ\begin{pmatrix} m & 0 \\ 0 & n \end{pmatrix}\circ\delta$ integral over $\mathbb{Z}[j] = \mathbb{Z}[j\circ\delta]$,

and (replacing τ by $\delta(\tau)$) in turn equivalent to $j\circ\begin{pmatrix} m & 0 \\ 0 & n \end{pmatrix}$ integral over $\mathbb{Z}[j]$. But $j\circ\begin{pmatrix} m & 0 \\ 0 & n \end{pmatrix}(\tau) = j(\frac{m}{n}\tau)$ with an integer m/n. By successive multiplications by prime numbers p we see that we are reduced to showing that $j(p\tau)$ is integral over $\mathbb{Z}[j(\tau)]$ for all primes p (and τ in the upper half plane). Let j_p be the function on the upper half plane defined by $j_p(\tau) = j(p\tau)$. We shall prove that the function j_p is integral over the ring of functions $\mathbb{Z}[j]$ (transcendental point of view).

Second step : $L_{p\tau}$ is a sublattice of L_τ of index p. Our guess (!) is that the conjugate algebraic numbers to $j_p(\tau) = j(p\tau)$ are the invariants of the other sublattices of index p in $L = L_\tau$. There is a lemma.

(3.11) Lemma 6. a) The lattices of index p in $\mathbb{Z}^2 = \langle e_1, e_2 \rangle$ are

$L_\nu = \langle e_1 + \nu e_2, pe_2 \rangle = \mathbb{Z}^2\cdot\alpha_\nu$ with $\alpha_\nu = \begin{pmatrix} 1 & \nu \\ 0 & p \end{pmatrix}$ for $0 \leq \nu < p$

and $L_p = \langle pe_1, e_2 \rangle = \mathbb{Z}^2\cdot\alpha_p$ with $\alpha_p = \begin{pmatrix} p & 0 \\ 0 & 1 \end{pmatrix}$. In particular, there are p + 1 lattices of index p in $\mathbb{Z}^2$.

b) Let $\Gamma = SL_2(\mathbb{Z})$, and define the matrices α_ν $(0 \leq \nu \leq p)$ as in a). Then we have the disjoint coset decomposition

$$\Gamma \begin{pmatrix} p & 0 \\ 0 & 1 \end{pmatrix} \Gamma = \coprod_{\nu=0}^{p} \Gamma\alpha_\nu \quad (\coprod : \text{disjoint union}) .$$

These are just two ways of expressing the same result as we shall see. Let Γ operate at right in $\mathbb{Z}^2$ (we consider elements of $\mathbb{Z}^2$ as line vectors, and we multiply them at right by square matrices, getting again line vectors). The mapping $\alpha \mapsto \mathbb{Z}^2\cdot\alpha$ gives a one-to-one correspondence between left cosets $\Gamma\alpha$ and sublattices $L = \mathbb{Z}^2\cdot\alpha$ of $\mathbb{Z}^2$. If we restrict this correspondence to matrices of determinant p, we shall get exactly those lattices L which are of index p in $\mathbb{Z}^2$. By elementary divisor theorem, every matrix of determinant p is in the double coset $\Gamma\begin{pmatrix} p & 0 \\ 0 & 1\end{pmatrix}\Gamma$. This explains the equivalence of the two assertions a) and b) of the lemma. Let us prove a). Let L be an arbitrary lattice of index p in $\mathbb{Z}^2$. Every element of $\mathbb{Z}^2/L$ is annihilated by p, so that $pe \in L$ for every $e \in \mathbb{Z}^2$. This proves that L lies between $\mathbb{Z}^2$ and $p\cdot\mathbb{Z}^2$: $p\cdot\mathbb{Z}^2 \subset L \subset \mathbb{Z}^2$, and more precisely, the lattices of index p correspond to the subgroups of index p in $\mathbb{Z}^2/p\cdot\mathbb{Z}^2$. Because $p\cdot\mathbb{Z}^2$ is of index p^2, these subgroups are of order p and are the lines (over $\mathbb{F}_p = \mathbb{Z}/p\cdot\mathbb{Z}$) through the origin in $\mathbb{F}_p \times \mathbb{F}_p$. These are the points of $\mathbb{P}^1(\mathbb{F}_p)$ and there are $p+1$ points on this projective line. Coming back to the lattices, we see that the

$$L_\nu = \langle e_1 + \nu e_2, pe_2\rangle = \mathbb{Z}^2\cdot\begin{pmatrix} 1 & \nu \\ 0 & p\end{pmatrix} \qquad (0 \leqslant \nu \leqslant p-1)$$

correspond to the non-vertical lines (through the origin) in $\mathbb{F}_p \times \mathbb{F}_p$, and

$$L_p = \langle pe_1, e_2\rangle = \mathbb{Z}^2\cdot\begin{pmatrix} p & 0 \\ 0 & 1\end{pmatrix} \text{ corresponds to the vertical one.}$$

This proves the lemma, and so we put $j_\nu(\tau) = j\cdot\alpha_\nu(\tau) = j(\frac{\tau+\nu}{p})$ (for $0 \leqslant \nu \leqslant p-1$) and form the so called <u>modular equation</u> of degree p

$$F_p(j,X) = \prod_{\nu=0}^{p} (X - j_\nu) = X^{p+1} + \sum_{n=0}^{p} \sigma_n X^n . \tag{3.12}$$

The $\sigma_n = \sigma_n(j_0,\ldots,j_p)$ are the elementary symmetric functions in the functions j_ν .

<u>Third step</u> : $F_p(j,X) \in \mathbb{C}[j][X] = \mathbb{C}[j,X]$.

The functions σ_n are indeed holomorphic functions over the upper half-plane and are invariant $\sigma_n\cdot\gamma = \sigma_n$ $(\gamma \in \Gamma)$ because by lemma 6b) right multiplication by γ permutes the left cosets $\Gamma\alpha_\nu$ and thus permutes the j_ν . Because the j_ν have a reasonable growth, say $|j_\nu(z)| \prec \exp(\Lambda_\nu \mathrm{Im}(z))$ for $\mathrm{Im}(z) \to \infty$, the same will be true for the functions σ_n . These functions σ_n are thus <u>holomorphic</u> modular functions, and by (I.3.11) are polynomials in j : $\sigma_n \in \mathbb{C}[j]$.

<u>Fourth step</u> : Let $P_n(j) = \sigma_n = \sum_{k \geq -N} a_k q^k$ be the Fourier expansion of σ_n (with $q = \underline{e}(\tau)$ and $N = N(n)$ equal to the degree of P_n). Then we show that the coefficients $a_k = a_k(n)$ are rational integers.

The functions j_ν have Fourier expansions in powers of $q^{1/p} = \underline{e}(\tau/p)$. Explicitely, if we introduce the primitive $p^{\underline{th}}$ root of one $\zeta = \underline{e}(1/p)$, we have

$$j_\nu(\tau) = j(\frac{\tau+\nu}{p}) = \underline{e}(-\frac{\tau+\nu}{p}) + \sum_{n\geq 0} c(n)\underline{e}(n\frac{\tau+\nu}{p}) =$$

$$(3.13) \qquad = \zeta^{-\nu}q^{-1/p} + \sum_{n\geq 0} c(n)\zeta^{\nu n}q^{n/p} \qquad (0 \leq \nu \leq p-1) \ ,$$

$$j_p(\tau) = j(p\tau) = q^{-p} + \sum_{n\geq 0} c(n)q^{pn} \qquad ,$$

with integers $c(n) \in \mathbb{Z}$ (the coefficients of the q-expansion of j are integers (I.4.4)). This shows that the coefficients $a_k = a_k(n)$ are in the ring $\mathbb{Z}[\zeta]$. But if $\rho \in \mathrm{Gal}(\mathbb{Q}(\zeta)/\mathbb{Q})$, then $\zeta^\rho = \zeta^r$ for some integer r, $1 \leq r \leq p-1$ and applying the automorphism ρ coefficient-wise to the q-expansions permutes the j_ν $(1 \leq \nu \leq p-1)$ and leaves j_0 as well as j_p fixed. The symmetric combinations will thus have invariant coefficients under all automorphisms in question and be in $\mathbb{Q} \cap \mathbb{Z}[\zeta] = \mathbb{Z}$ (simply because $\mathbb{Z}[\zeta]$ is contained in - in fact equal to - the ring of integers of $\mathbb{Q}(\zeta)$) .

<u>Fifth (final) step</u> : It is given by the following q-<u>expansions principle</u>, which will imply in particular $P_n(j) \in \mathbb{Z}[j]$, hence $F_p(j,X) \in \mathbb{Z}[j,X]$.

(3.14) <u>Lemma 7</u>. <u>Let</u> $P(j) = \sum_{k \geq -N} a_k q^k$ <u>be the</u> q-<u>expansion of a polynomial in the</u> j <u>function</u> (<u>with complex coefficients</u>). <u>Then the subgroup of</u> $\mathbb{C}$ <u>generated by the coefficients of</u> P <u>is equal to the group generated by the Fourier coefficients</u> a_k $(k \geq -N)$.

This lemma is trivial if N = 0 (P is a constant), so we use induction on N and suppose that it is already proved for N-1 instead of N.

Because j has a simple pole with residue one (in $q = 0$), N must be the degree of P and the coefficient of j^N in P must be a_{-N}. Consider

$$P(j) - a_{-N}j^N = \sum_{k\geq -N} a_k q^k - a_{-N}(q^{-1} + \sum_{k\geq 0} c(k)q^k)^N =$$

$$= \sum_{k\geq -(N-1)} b_k q^k \quad .$$

By induction hypothesis, the coefficients of $P(j) - a_{-N}j^N$ generate the group $\langle b_k\rangle = \bigoplus_{k\geq -(N-1)} \mathbb{Z}b_k$, so the coefficients of P(j) generate the group

$$\langle b_k\rangle + \mathbb{Z}a_{-N} \quad .$$

But this group is $\langle a_k\rangle_{k\geq -N}$ (because it contains a_{-N} and the c(k) are integers).

Thus theorem 2 is completely proved, and if we write j^* instead of the indeterminate X, $F_p(j,j^*) \in \mathbb{Z}[j,j^*]$ and the roots of $F_p(j,j^*) = 0$ are precisely the $j^* = j_\nu$ $(0\leq \nu\leq p)$. Because these roots are permuted transitively (lemma 6b) under the automorphisms $\tau \longmapsto \alpha(\tau)$ of the upper half-plane, F_p is irreducible as polynomial in j^* with coefficients in $\mathbb{Z}[j]$ (or $\mathbb{Z}(j)$ by Gauss' lemma, or still $\mathbb{C}(j)$).
We have identically

$$F_p(j(\tau),j(p\tau)) = 0 \qquad (\mathrm{Im}(\tau) > 0) ,$$

hence, dividing by p, also identically

$$F_p(j(\tau/p),j(\tau)) = 0 \quad .$$

This shows that the two polynomials $F_p(j,j^*)$, $F_p(j^*,j) \in \mathbb{C}(j)[j^*]$ have $j^* = j_0$ as common root, and by irreducibility of the polynomial

$$F_p(j,j^*) = (j^*)^{p+1} + \sigma_p(j^*)^p + \ldots ,$$

there must exist a polynomial $P(j,j^*) \in \mathbb{Z}[j,j^*]$ such that

$$F_p(j^*,j) = P(j,j^*)F_p(j,j^*) \quad .$$

Iterating this procedure of inversion will give

$$F_p(j,j^*) = P(j^*,j)F_p(j^*,j) = P(j^*,j)P(j,j^*)F_p(j,j^*) ,$$

and hence $P(j^*,j)P(j,j^*) = 1$ in $\mathbb{Z}[j,j^*]$: $P(j^*,j) = P(j,j^*) = \pm 1$.

If $P(j,j^*) = -1$, we have $F_p(j^*,j) = -F_p(j,j^*)$ and giving the value $j^* = j$, we would get identically $F_p(j,j) = 0$, so that $F_p(j,j^*)$ would have the root $j^* = j$ and be divisible by $j^* - j$, a contradiction to the irreducibility of $F_p(j,j^*)$ in $\mathbb{Z}[j][j^*]$. This proves that <u>the modular polynomial</u> F_p <u>is symmetrical in</u> j <u>and</u> j^*.

These polynomials can be very difficult to determine explicitely as the case $p = 2$ already shows (Bateman t.3,p.25)

$$F_2(j,j^*) = j^3 + j^{*3} - (jj^*)^2 + 3^4 5^3 4027\, jj^* + 2^4 3\cdot 31\, jj^*(j + j^*)$$
$$- 2^4 3^4 5^3 (j^2 + j^{*2}) + 2^8 3^7 5^6 (j + j^*) - 2^{12} 3^9 5^9 \quad .$$

It can be shown however, that

$$F_p(j,j^*) \equiv (j - j^{*p})(j^p - j^*) \quad \text{mod } p \quad .$$

(3.15) <u>Theorem 3</u>. <u>When</u> L <u>is a singular lattice</u>, $j(L)$ <u>is an algebraic integer</u>.

<u>Proof</u>. We observe that L is isogenic to the ring of integers in the quadratic field $\mathrm{End}(L_{\mathbb{Q}}) = K$, so by transitivity of the notion of integrality, it is sufficient to prove that $j(\mathfrak{O}_K)$ itself is an integer. There exists an element $a \in K$ having a prime norm $N(a) = p$. For example, take a prime p which is not inert in K (i.e. which does not generate a prime ideal of $\mathfrak{O}_K$), $p\mathfrak{O}_K = \mathfrak{p}\mathfrak{q}$ with principal ideals $\mathfrak{p}$, $\mathfrak{q}$, necessarily of norm p. Then $N(\mathfrak{p}) = \operatorname{Inf}_{a\in\mathfrak{p}} N(a)$ gives an element of prime norm p. This shows that $\mathfrak{O}_K$ is isomorphic to a sublattice (namely $a\mathfrak{O}_K$) of index p in $\mathfrak{O}_K$:

$$p = N(a) = \mathrm{Card}(\mathfrak{O}_K/a\mathfrak{O}_K) = [\mathfrak{O}_K : a\mathfrak{O}_K] \; .$$

By definition of the modular polynomial, we have $F_p(j(\mathfrak{O}_K),j(\mathfrak{O}_K)) = 0$. But the polynomial $F_p(j,j) \in \mathbb{Z}[j]$ has leading term -1 as the q-expansions of the $j - j_\nu$ show :

$$(j - j_\nu)(\tau) = q^{-1} + \text{higher order terms} \quad (\text{for } 0 \leqslant \nu \leqslant p-1) \; ,$$
$$(j - j_p)(\tau) = -q^{-p} + \text{higher order terms}$$

(these expansions are to be regarded as power series in $q^{1/p}$), and so

$$F_p(j,j) = \prod_{\nu=0}^{p} (j - j_\nu) = \underbrace{q^{-1}\cdots q^{-1}}_{p\ \text{terms}}(-1/q^p) + \text{higher order terms}$$

$$= -q^{-2p} + \text{higher order terms} = -j^{2p} + \ldots \quad .$$

This proves that $-F_p(j,j)$ is a unitary polynomial of degree 2p in j giving an integral equation of dependance of $j(\mathfrak{o}_K)$ over $\mathbb{Z}$.

Now we turn to the p-adic proof of Th.3, not using Th.2 (but using Th.1 which is comparatively very simple). For that, we shall prove

(3.16) Theorem 4. Let L be a finite algebraic extension of the p-adic field $\mathbb{Q}_p$, $q \in L^\times$ such that $|q| < 1$ and $E_q = E_q(\bar{L})$ be the corresponding elliptic curve of Tate. Then $\mathrm{End}(E_q) = \mathbb{Z}$.

Proof. (We are using L instead of K for the p-adic field, keeping K for $\mathrm{End}(E_q) \otimes \mathbb{Q}$.) First step : For every algebraic extension L' of L we have a parametrization

$$\varphi : L'^\times \longrightarrow E_q(L') \subset \mathbb{P}^2(L')$$

$$x \longmapsto (P(x), DP(x), 1) \quad \text{if } x \notin q^{\mathbb{Z}} \quad ,$$

giving isomorphisms

$$\bar{\varphi} : L'^\times/q^{\mathbb{Z}} \xrightarrow{\sim} E_q(L')$$

(observe that the theorem is certainly true for finite algebraic extensions L' of L by our treatment of Tate's curves in chapter II , hence also for every algebraic extension of L, because such an extension is union of finite algebraic ones). Now φ being composite of two homomorphisms is a homomorphism itself, and if we denote by + the group law of E_q we have

$$\varphi(xy) = \varphi(x) + \varphi(y) \qquad (x,y \in L'^\times) \ .$$

Let us take now an L-automorphism σ of an algebraic closure $\bar{L}$ of L . This automorphism is automatically continuous on $\bar{L}$ for the topology deduced by the unique extension of the absolute value of L (L is complete). In particular σ transforms a sequence tending to 0 in another sequence tending to 0, hence a convergent series in another convergent series (the conjugate of the sum can be computed by taking the sum of the conjugates). Because the p-adic functions P and DP have their coefficients in the field L ($q \in L$ and $P(X) = P(X:q)$ is the analogue of the q-expansion of the Weierstrass function) , we shall have

$$P(x^\sigma) = P(x)^\sigma \quad , \quad DP(x^\sigma) = DP(x)^\sigma$$

hence also $\varphi(x^\sigma) = \varphi(x)^\sigma$. This proves that the <u>parametrization is compatible with the action of</u> $\mathrm{Gal}(\bar{L}/L)$ <u>on</u> $\bar{L}^\times$ <u>on one hand, and on</u> $E_q(\bar{L}) \subset \mathbb{P}^2(\bar{L})$ <u>on the other</u> (in other words, φ commutes with the automorphisms $\sigma \in \mathrm{Gal}(\bar{L}/L)$) .

<u>Second step</u> : Let $T_\ell = T_\ell(E_q)$ denote Tate's module for the prime ℓ. We shall show that if ℓ does not divide $\mathrm{ord}_L(q)$ $(= -\mathrm{ord}_L j(E_q))$, then <u>the image</u> $\widetilde{G}_\ell$ <u>of</u> $\mathrm{Gal}(\bar{L}/L)$ <u>in</u> $\mathrm{Aut}(T_\ell/\ell T_\ell) = \mathrm{Aut}(t_\ell(E_q))$ <u>contains a unipotent operator with matrix</u> $\begin{pmatrix} 1 & 1 \\ 0 & 1 \end{pmatrix}$ <u>in a suitable basis</u> (<u>over</u> $\mathbb{F}_\ell$) <u>of this space</u>. We denote by L' the cyclotomic extension of L obtained by adjoining a primitive ℓ^{th} root of 1 to L (hence L' contains all roots of 1 of order ℓ). Because $[L':L]$ is a divisor of $\ell - 1$, hence prime to ℓ , we shall still have ℓ prime to $\mathrm{ord}_{L'}(q)$ $(= e\cdot\mathrm{ord}_L q$ with the ramification index $e = e(L'/L)$ dividing the degree of $L'/L)$.

In particular, q has still no ℓ^{th} root in L'. Choose an ℓth root $q^{1/\ell} \in \bar{L}$ and an automorphism $\sigma \in \mathrm{Gal}(\bar{L}/L') \subset \mathrm{Gal}(\bar{L}/L)$ such that $\sigma(q^{1/\ell}) \neq q^{1/\ell}$. Necessarily $\sigma(q^{1/\ell})/q^{1/\ell} = \zeta$ is a primitive ℓ^{th} root of 1 (hence in L'). Put now $e_1 = \varphi(\zeta)$, $e_2 = \varphi(q^{1/\ell})$. This is a basis of $t_\ell(E_q)$ and in this basis, the action of the automorphism σ is described by a unipotent matrix as asserted :

$$e_1^\sigma = \varphi(\zeta)^\sigma = \varphi(\zeta^\sigma) = \varphi(\zeta) = e_1 \quad ,$$
$$e_2^\sigma = \varphi(q^{1/\ell})^\sigma = \varphi(\sigma(q^{1/\ell})) = \varphi(\zeta\cdot q^{1/\ell}) = \varphi(\zeta) + \varphi(q^{1/\ell}) = e_1 + e_2 \,.$$

Third step : There is a finite algebraic extension L'' of L so that every endomorphism $\alpha \in \mathrm{End}(E_q)$ is defined over L''. There is nothing to prove if $\mathrm{End}(E_q) = \mathbb{Z}$, so we suppose that $\mathrm{End}(E_q) = \mathcal{O}_f = \mathbb{Z} + \mathbb{Z}\alpha$ is an order in an imaginary quadratic field. The assertion will be proved if we only prove that the endomorphism α is defined over a finite algebraic extension of L (multiplication by an integer in E_q is defined over L). But we can embed L into the complex field $\mathbb{C}$ (algebraically closed and having a transcendence basis over the prime field $\mathbb{Q}$ having the power of the continuum). Then for every automorphism $\sigma \in \mathrm{Aut}(\mathbb{C}/L)$, α^σ is still an automorphism of E_q . This proves that the set of conjugates $\{\alpha^\sigma\}$ is at most denumerable. This proves that α is algebraic, defined over a finite algebraic extension of L. As consequence, if $t \in t(E_q)$ is a division point on E_q, if $\alpha \in \mathrm{End}(E_q)$ is defined over L'' and $\sigma \in \mathrm{Gal}(L''/L)$, we shall have $\alpha(t^\sigma) = \alpha(t)^\sigma$. This shows that the ℓ-adic representations of $\mathrm{End}(E_q)$ and $\mathrm{Gal}(\bar{L}/L'')$ in $T_\ell(E_q)$ commute.

Fourth step. We conclude the proof by contradiction. Let us replace L by a finite algebraic extension (still denoted by L) over which every $\alpha \in \mathrm{End}(E_q)$ is defined. We have proved that for all prime integers ℓ not dividing $\mathrm{ord}_L(q)$, there is an element $\sigma \in \mathrm{Gal}(\bar{L}/L)$ having representing matrix $\begin{pmatrix}1 & 1\\ 0 & 1\end{pmatrix}$ for its action on $t_\ell(E_q) = T_\ell/\ell T_\ell$. Since all elements of $\mathrm{End}(E_q)$ must have representations in this space commuting with this unipotent matrix, they must themselves be in upper triangular form with two equal characteristic values :

$$\begin{pmatrix}a & a+b\\ c & c+d\end{pmatrix} = \begin{pmatrix}a & b\\ c & d\end{pmatrix}\begin{pmatrix}1 & 1\\ 0 & 1\end{pmatrix} = \begin{pmatrix}1 & 1\\ 0 & 1\end{pmatrix}\begin{pmatrix}a & b\\ c & d\end{pmatrix} = \begin{pmatrix}a+c & b+d\\ c & d\end{pmatrix}$$

implies

$$c = 0 \quad \text{and } a = d \quad .$$

Hence all elements of the image of $\mathrm{End}(E_q)$ in $\mathrm{End}(t_\ell(E_q))$ would have equal eigenvalues. This would lead to a contradiction if $\mathrm{End}(E_q)$

generated a quadratic field $K = \mathbb{Q}(\sqrt{-d})$, as soon as ℓ would be taken prime to d .

(3.17) <u>Corollary</u>. <u>If</u> τ <u>is imaginary quadratic</u> (<u>with</u> $\mathrm{Im}(\tau) > 0$), <u>then</u> $j(\tau)$ <u>is an algebraic integer</u>.

<u>Proof</u>. By Th.1, $j(\tau)$ is algebraic, so that $\mathbb{C}/L_\tau$ is isomorphic to a non-singular cubic curve defined over a number field k. If $j(\tau)$ were not an integer, there would be a prime ideal $\mathfrak{p}$ of the ring of integers $\mathcal{O}_k$ of k appearing with a negative exponent in the decomposition of the principal fractional ideal generated by $j = j(\tau)$, hence a prime ideal $\mathfrak{p}$ with $\mathrm{ord}_{\mathfrak{p}}\, j < 0$. In the completed p-adic field $k_{\mathfrak{p}}$ of k, there would exist a unique $q \in k_{\mathfrak{p}}^{\times}$ with $|q| < 1$ and $j = j(q)$. Then our algebraic cubic would be isomorphic to Tate's curve E_q over $\bar{k}_{\mathfrak{p}}$ (or already over a quadratic extension of $k_{\mathfrak{p}}$). But we have seen that $\mathrm{End}(E_q) = \mathbb{Z}$ for all Tate's curves, contradicting our hypothesis that τ being imaginary quadratic, $\mathrm{End}(\mathbb{C}/L_\tau) \neq \mathbb{Z}$.

The third step in the proof of Th.4 above can be made much more explicit: if E is an elliptic curve defined over a field k (of characteristic 0), <u>then every endomorphism of</u> E <u>is defined over</u> kK, where K is the field generated by End(E). This is trivial if $\mathrm{End}(E) = \mathbb{Z}$. The other case will be treated now by using first kind differential forms on E. We have to review some notions.

Let V be an absolutely irreducible (projective plane) algebraic curve defined over the field k of characteristic zero. Thus $V = V_\Omega$ where Ω is an algebraically closed field containing k. If σ is an automorphism of Ω over k, then σ acts on the space $\mathrm{Der}(V) = \mathrm{Der}_\Omega(V)$ of derivations $D : \Omega(V) \longrightarrow \Omega(V)$ of the rational function field of V (trivial over Ω). In fact one defines $D^\sigma \in \mathrm{Der}(V)$ by the formula

(3.18) $$D^\sigma(f^\sigma) = D(f)^\sigma \quad \text{for } f \in \Omega(V).$$

(Here, as usual, we denote by f^σ the rational function on V having conjugate coefficients under σ in a representation of f as quotient P/Q of two polynomial functions on V : this definition of f^σ is independent of the choice of the representation P/Q because the ideal of definition of V is generated by polynomials with coefficients in k hence is invariant under σ.) If we recall that the derivation D_f with respect to a rational function is uniquely defined by the property that $D_f(f) = 1$ (we suppose that f is not constant), we see that

$$D_f^\sigma(f^\sigma) = D_f(f)^\sigma = 1 = D_{f^\sigma}(f^\sigma) ,$$

hence

$$D_f^\sigma = D_{f^\sigma} \quad . \tag{3.19}$$

Now if $\omega \in \mathrm{Diff}(V)$ is a differential form on V, i.e. an $\Omega(V)$-linear form over Der(V) (these two spaces have dimension 1 over $\Omega(V)$), we also define

$$\langle \omega^\sigma, D^\sigma \rangle = \langle \omega, D \rangle^\sigma \quad (\text{for } D \in \mathrm{Der}(V)). \tag{3.20}$$

In particular if $\omega = df$ ($f \in \Omega(V)$), we have

$$\langle (df)^\sigma, D^\sigma \rangle = \langle df, D \rangle^\sigma = D(f)^\sigma = D^\sigma(f^\sigma) = \langle d(f^\sigma), D^\sigma \rangle ,$$

hence $(df)^\sigma = d(f^\sigma)$ and more generally

$$\omega = g\,df \quad \text{implies} \quad \omega^\sigma = g^\sigma d(f^\sigma) \quad . \tag{3.21}$$

Let now $P \in V$ be a regular point, $\pi_P \in \Omega(V)$ a uniformizing variable at P. Because $\langle d\pi_P, D_{\pi_P} \rangle = 1$ every differential form ω on V can be written

$$\omega = \langle \omega, D_{\pi_P} \rangle \, d\pi_P \quad . \tag{3.22}$$

The proof of (2.1.c) shows that $\mathrm{ord}_{P^\sigma} = (\mathrm{ord}_P)^\sigma$ hence that we can choose $\pi_P^\sigma = \pi_{P^\sigma}$ as uniformizing variable at P^σ. In particular, $\langle \omega^\sigma, D_{\pi_{P^\sigma}} \rangle$ is regular at P (in R_P) whenever $\langle \omega, D_{\pi_P} \rangle$ is regular at P (the local ring R_{P^σ} at P^σ is R_P^σ) . This shows that if $\omega \in \mathcal{D}^1$ is a first kind differential form on V, its conjugate ω^σ will also be a first kind differential form on V.

If $\omega = \omega^{\sigma}$ for all $\sigma \in \mathrm{Aut}(\Omega/k)$, then ω is a k-rational differential form on V (take $f \in k(V)$ not constant, and write $\omega = g\,df$ with some function $g \in \Omega(V)$: $\omega^{\sigma} = \omega$ gives immediately $g^{\sigma} = g$ for all k-automorphisms σ of Ω).

Let now $\alpha : V \longrightarrow W$ be a rational mapping (when expressed in a coordinate system in W, the coordinates of α must be rational functions on V) between two curves (satisfying the same assumptions as the curve V of the last paragraph). Thus α defines a homomorphism

$$\alpha^* : \Omega(V) \longleftarrow \Omega(W) \ , \qquad f\circ\alpha \longleftarrow\!\!| \ f \ .$$

We suppose α not constant, so that $\alpha^* \neq 0$ will be <u>injective</u>. <u>Certain</u> derivations $D \in \mathrm{Der}(V)$ have a direct image under α. For example if $D = D_{f\circ\alpha}$ $(f \in \Omega(W))$ is a derivation with respect to a rational function in the image of α^* we can define $\alpha_*(D) \in \mathrm{Der}(W)$ by the following procedure. Noting that $D_{f\circ\alpha}$ sends $f\circ\alpha$ on 1 (by definition), hence leaves stable $\Omega(W)\circ\alpha \subset \Omega(V)$, we have

$$D_{f\circ\alpha}(g\circ\alpha) = h\circ\alpha \qquad h \in \Omega(W) \ ,$$

and $\alpha_*(D_{f\circ\alpha})$ is defined by $g \longmapsto h$. In a formula, this reads

(3.23) $$\alpha_*(D_{f\circ\alpha})(g)\circ\alpha = D_{f\circ\alpha}(g\circ\alpha) \ .$$

Then by "duality" we can define the inverse image (under α) of <u>any</u> differential form $\omega \in \mathrm{Diff}(W)$ by putting

(3.24) $$\langle \alpha^*(\omega), D_{f\circ\alpha} \rangle_V = \langle \omega, \alpha_* D_{f\circ\alpha} \rangle_W \circ \alpha$$

(we put indices V or W with the bilinear symbol $\langle \cdot , \cdot \rangle$ to make more precise the curve with respect to which it is defined). This formula characterizes $\alpha^*(\omega)$ completely with a single non-constant $f \in \Omega(W)$ because then $D_{f\circ\alpha} \neq 0$ and $\mathrm{Diff}(V)$ is one-dimensional over $\Omega(V)$. In particular if $\omega = df$ $(f \in \Omega(W))$, we get

$$\langle df, \alpha_* D_{f\circ\alpha} \rangle_W \circ \alpha = \alpha_*(D_{f\circ\alpha})(f)\circ\alpha = D_{f\circ\alpha}(f\circ\alpha) = 1,$$

whence $\alpha^*(df) = d(f\circ\alpha)$ and more generally it follows

(3.25) $$\alpha^*(g\,df) = (g\circ\alpha)\,d(f\circ\alpha) \quad (f,g \in \Omega(W)).$$

This formula justifies another notation for $\alpha^*(\omega)$, namely $\omega\circ\alpha$.
If σ is a k-automorphism of Ω, and $f\in\Omega(W)$ a rational function on W, it is obvious that $(f\circ\alpha)^\sigma = f^\sigma\circ\alpha^\sigma$ and from there using (3.25), that

(3.26) $$\alpha^*(\omega)^\sigma = (\omega\circ\alpha)^\sigma = \omega^\sigma\circ\alpha^\sigma = (\alpha^\sigma)^*(\omega^\sigma) \quad .$$

Let $P\in V$ be a regular point, $\alpha(P) = Q$ be a well-defined regular point on W (thus we suppose α regular at P : $\alpha^*(R_Q) = R_Q\circ\alpha \subset R_P$). If π_P and π_Q are uniformizing variables at P and Q respectively, we shall have $\pi_Q\circ\alpha\in R_P$ (and this implies $D_{\pi_P}(\pi_Q\circ\alpha)\in R_P$ by (II.3.10)). Then <u>if the differential form</u> $\omega\in \mathrm{Diff}(W)$ <u>is regular at</u> Q, $\alpha^*(\omega)$ <u>will be regular at</u> P. By (3.22) the hypothesis means that $\langle\omega,D_{\pi_Q}\rangle_W$ is regular at Q. Let us compute

$$\langle\alpha^*(\omega),D_{\pi_P}\rangle_V = \langle\alpha^*(\omega),D_{\pi_P}(\pi_Q\circ\alpha)D_{\pi_Q\circ\alpha}\rangle_V \quad ,$$

because

$$D_{\pi_P}/D_{\pi_Q\circ\alpha} = D_{\pi_P}(\pi_Q\circ\alpha)/D_{\pi_Q\circ\alpha}(\pi_Q\circ\alpha) = D_{\pi_P}(\pi_Q\circ\alpha) \ .$$

Hence

$$\langle\alpha^*(\omega),D_{\pi_P}\rangle_V = D_{\pi_P}(\pi_Q\circ\alpha)\Big[\big\langle\langle\omega,D_{\pi_Q}\rangle d\pi_Q \, , \alpha_* D_{\pi_Q\circ\alpha}\big\rangle_W\circ\alpha\Big] =$$

$$= D_{\pi_P}(\pi_Q\circ\alpha)\Big[\langle\omega,D_{\pi_Q}\rangle_W\circ\alpha\Big]\Big[\langle d\pi_Q, \alpha_* D_{\pi_Q\circ\alpha}\rangle_W\circ\alpha\Big] \ .$$

The first term in this product $D_{\pi_P}(\pi_Q\circ\alpha)$ is in R_P because α is regular at P. The second term $\langle\omega,D_{\pi_Q}\rangle_W\circ\alpha$ is also in R_P because ω is regular at Q and α regular at P. The last term is

$$\langle d\pi_Q, \alpha_* D_{\pi_Q\circ\alpha}\rangle_W\circ\alpha = \langle\alpha^* d\pi_Q, D_{\pi_Q\circ\alpha}\rangle_V = 1 \quad .$$

This proves the assertion. In consequence, if α is regular on the whole of V (non-singular) and $\omega\in\mathcal{D}^1$ is a first kind differential on W, then $\alpha^*(\omega)$ is a first kind differential on V.

(3.27) <u>Proposition</u>. <u>Let</u> E <u>be an elliptic curve in characteristic</u> 0 <u>admitting complex multiplications</u> : $\mathrm{End}(E)\neq\mathbb{Z}$. <u>Let</u> k ($\subset\mathbb{C}$) <u>be a field of definition of</u> E <u>and</u> K ($\subset\mathbb{C}$) <u>be the field</u> $\mathrm{End}(E)\otimes_{\mathbb{Z}}\mathbb{Q}$ <u>genera-</u>

<u>ted by the complex multiplications of</u> E. <u>Then every endomorphism of</u> E <u>is defined over</u> kK.

<u>Proof</u>. We use the representation $\alpha \mapsto \alpha^*$ of the ring of endomorphisms of E in the space of first kind differential forms over E. Because this space is of dimension one over $\mathbb{C}$ (II.3.16) α^* must be a homothety by a well-defined complex number a_α :

$$\begin{array}{ccc} \mathrm{End}(E) & \longrightarrow & \mathrm{End}(\mathfrak{D}^1_{\mathbb{C}}) \cong \mathbb{C} \text{ (canonical isomorphism)} \\ \alpha & \longmapsto & a_\alpha . \end{array}$$

Let now $\omega \in \mathfrak{D}^1_k$ be a k-rational first kind differential form on E, and σ an automorphism of $\mathbb{C}$ over kK. We have

$$\alpha^*(\omega)^\sigma = \begin{cases} (\omega \circ \alpha)^\sigma = \omega \circ \alpha^\sigma = a_{\alpha^\sigma} \cdot \omega \\ (a_\alpha \omega)^\sigma = a_\alpha^\sigma \cdot \omega = a_\alpha \cdot \omega \end{cases},$$

hence $a_{\alpha^\sigma} = a_\alpha$. But the above isomorphism extends to a (normalized) embedding $\mathrm{End}(E) \otimes_{\mathbb{Z}} \mathbb{Q} \longrightarrow \mathbb{C}$ (with image K), and in particular is injective. From $a_{\alpha^\sigma} = a_\alpha$ we derive thus $\alpha^\sigma = \alpha$ and this shows that α is defined over kK. (The representation of End(E) in $\mathfrak{D}^1_{\mathbb{C}}$ is simpler to explain in transcendental terms : if $\mathbb{C}/L \xrightarrow{\sim} E$ induces the isomorphism $\mathrm{End}(L) \xrightarrow{\sim} \mathrm{End}(E)$ and dz is the differential form of first kind applied onto ω , then $a \in \mathrm{End}(L)$ induces the homothety $dz \longmapsto d(az) = a\,dz$.)

4. Division points in characteristic p

This section is descriptive in character, its goal being to indicate a possible approach to division points in characteristic $p \neq 0$ (we shall also assume $p \neq 2,3$ to be able to work with Weierstrass normal forms, although results are true quite generally).

Let E be an elliptic curve of (affine) equation $q^2 = 4p^3 - g_2p - g_3$ $(\Delta = g_2^3 - 27g_3^2 \neq 0)$. We know that it is uniformized by the Weierstrass functions $p = p(z) = \wp(z)$ and $q = p' = \wp'(z)$. We are trying to express $p_n = p(nz)$ and $q_n = p'(nz)$ as rational functions of p and q. It is easy to do so for n = 2 noting that the addition formula (I.1.14) for $u \to v = z$ leads to the duplication formula (I.1.15)

(4.1) $$p_2 = -2p + \tfrac{1}{4}p''^2/p'^2 \quad .$$

By derivation of the identity $p'^2 = 4p^3 - g_2p - g_3$ we get

$$2p'p'' = 12p^2p' - g_2p' \quad , \qquad p'' = 6p^2 - \tfrac{1}{2}g_2 \quad .$$

To find an expression for q_2, we have to find the equation $q = mp + h$ of the line tangent to E at (p,q). After some elementary computations, we find $m = p''/p'$, $h = (p'^2 - pp'')/p'$ so that

(4.2) $$q_2 = -mp_2 - h = 3pp''/p' - \tfrac{1}{4}p''^3/p'^3 - p' \quad .$$

To find an expression for p_3 we use the addition formula

$$p_3 = -p_2 - p + \tfrac{1}{4}(q_2 - q)^2/(p_2 - p)^2 \quad .$$

By (4.2) $q_2 - q = q_2 - p' = (12pp'^2p'' - p''^3 - 8p'^4)/(4p'^3)$

and $p_2 - p = (p''^2 - 12pp'^2)/(4p'^2)$ so that

(4.3) $$\frac{q_2 - q}{p_2 - p} = \frac{1}{p'} \frac{12pp'^2p'' - p''^3 - 8p'^4}{p''^2 - 12pp'^2} \quad .$$

From this, it is easy to find

(4.4) $$p_3 = p - \tfrac{1}{4}\frac{1}{p'^2}\, 8p'^4(24pp'^2p'' - 2p''^3 - 8p'^4)/(p''^2 - 12pp'^2)^2 =$$

$$= p - 4p'^2(12pp'^2p'' - p''^3 - 4p'^4)/(p''^2 - 12pp'^2)^2 \quad .$$

To simplify the computations somewhat, put $g_2 = -4a$, $g_3 = -4b$, so that

$$p'' = 6p^2 + 2a = 2(3p^2 + a) \quad . \tag{4.5}$$

Then we see after some more computations

$$p_3 = p - \frac{p'^2(2p^6 + 10ap^4 + 40bp^3 - 10a^2p^2 - 8abp - 2a^3 - 16b^2)}{(3p^4 + 6ap^2 + 12bp - a^2)^2} \quad . \tag{4.6}$$

The equation of the curve E is now $q^2 = 4p^3 + 4ap + 4b$ so we put $x = p$ and $y = \frac{1}{2}q = \frac{1}{2}p'$ so as to have the equation

$$y^2 = x^3 + ax + b \quad . \tag{4.7}$$

Let us also define $(x_n, y_n) = n\cdot(x,y)$ so that $x_1 = x = p = p_1$, $x_n = p_n$ and $y_1 = y = \frac{1}{2}p'$, $y_n = \frac{1}{2}q_n$. Then (4.1) gives

$$x_2 = x - (3x^4 + 6ax^2 + 12bx - a^2)/(4y^2) \quad , \tag{4.8}$$

and (4.6) gives

$$x_3 = x - \frac{8y^2(x^6 + 5ax^4 + 20bx^3 - 5a^2x^2 - 4abx - 8b^2 - a^3)}{(3x^4 + 6ax^2 + 12bx - a^2)^2} \quad . \tag{4.9}$$

Now it can be proved by induction that

$$\begin{aligned} x_n &= x - \psi_{n-1}\psi_{n+1}/\psi_n^2 \\ y_n &= (\psi_{n+2}\psi_{n-1}^2 - \psi_{n-2}\psi_{n+1}^2)/(4y\psi_n^3) \end{aligned} \tag{4.10}$$

with polynomials ψ_n defined recursively by

$$\begin{aligned} \psi_{2n} &= \psi_n(\psi_{n+2}\psi_{n-1}^2 - \psi_{n-2}\psi_{n+1}^2)/(2y) \quad , \\ \psi_{2n+1} &= \psi_{n+2}\psi_n^3 - \psi_{n-1}\psi_{n+1}^3 \quad . \end{aligned} \tag{4.11}$$

To start the induction in (4.11) we need the first values given by (4.8) and (4.9) :

$$\begin{aligned} \psi_1 &= 1 \quad , \\ \psi_2 &= 2y \quad , \\ \psi_3 &= 3x^4 + 6ax^2 + 12bx - a^2 \quad , \\ \psi_4 &= 4y(x^6 + 5ax^4 + 20bx^3 - 5a^2x^2 - 4abx - 8b^2 - a^3) \quad . \end{aligned} \tag{4.12}$$

These polynomials are universal polynomials in x, a, b

$$\psi_{2n}/y \ , \ \psi_{2n+1} \in \mathbb{Z}[x,a,b]$$

with integral coefficients and have the following properties (which can be verified by induction...) :

(4.13) a) $x\psi_n^2 - \psi_{n-1}\psi_{n+1}$ is a polynomial in x of degree n^2 and leading coefficient 1 (we have used $y^2 = x^3 + ax + b$),

b) ψ_n^2 is a polynomial in x of degree n^2-1 ,

c) $x\psi_n^2 - \psi_{n-1}\psi_{n+1}$ and ψ_n^2 are prime to each other (considered as polynomials in x) and stay so even after reduction mod p (p being a prime $\neq 2,3$).

(For this last property, cf. J.W.S. Cassels: A note on the division values of $\wp(u)$, Proc. Cambridge Phil. Soc., 45 ,(1949),pp.167-172 .) These properties (which are classical in characteristic 0) are purely algebraic and have the following consequences:

(4.14) Theorem. Let E be the elliptic curve given by (4.7) with coefficients a,b in an algebraically closed field k of characteristic p, and denote by $n_E = n1_E : E \longrightarrow E$ the algebraic homomorphism $(x,y) \longmapsto n(x,y) = (x_n,y_n)$. Then

$$[k(E) : k(E)\cdot n_E] = n^2 \quad .$$

Proof (for $p \neq 2,3$). We start by proving $[k(x) : k(x_n)] = n^2$, and for that purpose we use Lüroth's theorem giving the degree of the left-hand side as the degree of the rational fraction x_n :

$$\deg(x_n) = \deg\ (x\psi_n^2 - \psi_{n-1}\psi_{n+1})/\psi_n^2 =$$

$$= \mathrm{Max}(\deg(x\psi_n^2 - \psi_{n-1}\psi_{n+1}),\deg(\psi_n^2))$$

by property c) above (we consider the reduced polynomials mod p as polynomials in x and coefficients in k). From property b) we infer that $\deg(\psi_n^2) \leqslant n^2 - 1$ (reduction mod p can lower the degree!) and thus finally $[k(x) : k(x_n)] = \deg_x(x_n) = n^2$, by property a) above.

Then we consider the following diagram of fields

$$\begin{array}{ccc} k(x) & \text{---} & k(x,y) = k(E) \\ {\scriptstyle n^2} \Big| & & \Big| \\ k(x_n) & \text{---} & k(x_n,y_n) = k(E)\cdot n_E \end{array}$$

Now the automorphism $(x,y) \longmapsto (x,-y)$ of $k(E)$ over $k(x)$ induces the non-trivial automorphism $(x_n,y_n) \longmapsto (x_n,-y_n)$ of $k(E)\cdot n_E$ over $k(x_n)$ (use (4.10) to check this, considering separately the cases n odd and n even). This proves that $k(x)$ and $k(x_n,y_n)$ are linearly disjoint over $k(x_n)$ and consequently

$$[k(x,y) : k(x_n,y_n)] = [k(x) : k(x_n)] = n^2 \quad .$$

(There is fortunately a proof of this basic theorem avoiding completely the computations relative to the polynomials ψ_n, but it uses intersection theory and is less elementary.)

If $\alpha : E \longrightarrow E$ is a homomorphism, it is useful to define the <u>degree</u> (resp. <u>separable</u>, <u>inseparable degree</u>) of α by

(4.15) $$\deg(\alpha) = [k(E) : k(E)\cdot\alpha] = \deg_s(\alpha)\cdot\deg_i(\alpha) \ ,$$

$$\deg_s(\alpha) = [k(E) : k(E)\cdot\alpha]_s \ ,$$

$$\deg_i(\alpha) = [k(E) : k(E)\cdot\alpha]_i \ ,$$

and to say that α is <u>separable</u> if $k(E)$ is separable over $k(E)\cdot\alpha$.

(4.16) <u>Corollary 1</u>. <u>The degree of</u> n_E <u>is</u> n^2 <u>and if</u> n <u>is prime to</u> p, n_E <u>is separable</u> : $(n,p) = 1 \Longrightarrow \deg(n_E) = \deg_s(n_E) = n^2$.

<u>Proof</u>. If $(n,p) = 1$, n^2 is not divisible by p and $\deg_i(n_E)$ which must be a power of p (and divide n^2) must be 1.

(4.17) <u>Corollary 2</u>. <u>If</u> n <u>is prime to</u> p, $t_n(E) \cong (\mathbb{Z}/n\mathbb{Z})^2$ <u>and so for any prime</u> $\ell \neq p$, $T_\ell(E)$ <u>is a free</u> $\mathbb{Z}_\ell$<u>-module of rank two and</u> $V_\ell(E)$ <u>is a vector space of dimension two over</u> $\mathbb{Q}_\ell$.

<u>Proof</u>. We look at the following diagram of algebraic curves and morphisms :

$$(4.18)\qquad \begin{array}{ccccc} x & \mathbb{P}^1 & \longleftarrow & E & (x,y) \\ \downarrow & \downarrow{\scriptstyle x_n} & & \downarrow{\scriptstyle n_E} & \downarrow \\ x_n(x) & \mathbb{P}^1 & \longleftarrow & E & (x_n,y_n) \end{array}$$

where the horizontal maps are the first projections (double sheeted coverings with ramification points of index two at division points of order two) and the left vertical map is the rational function $x_n = x_n(x)$. It is obvious that if $a \in \mathbb{P}^1$ is fixed, $x_n(x) = a$ has at most n^2 solutions $x \in \mathbb{P}^1$. Let us prove that it has indeed n^2 distinct solutions except for finitely many exceptional values of a. Let us write $x_n(x) = u(x)/v(x)$ with the polynomials given by (4.10), $(u,v) = 1$. If the polynomial $u(x) - av(x)$ has a multiple root ξ, then

$$u(x) - av(x) = (x-\xi)^2 r(x) \quad ,$$

$$u(x)/v(x) = a + (x-\xi)^2 r(x)/v(x) \quad ,$$

$$(u(x)/v(x))' = (x-\xi)\, s(x)$$

with a rational function $s(x)$ (having v^2 as denominator, and v does not have the root ξ), so that $(u/v)'$ still vanishes at ξ. If we discard the finite number of values where $v(x) = 0$, this implies that ξ is a root of $u'v-uv'$. This proves the assertion for all values of a except those $a = u(\lambda)/v(\lambda)$ where λ is a root of $u'v-uv'$. (Observe that $u'v - uv'$ is not identically 0 because $u'v = uv'$ would imply that $u'v$ is divisible by u hence u' divisible by u because $(u,v) = 1$, hence $u' = 0$, deg(u) divisible by p, which is not the case.) To conclude the proof of the corollary, it is sufficient to prove that the kernel of n_E consists of $m = n^2$ distinct points: if $n = \ell$ is a prime number $\neq p$, $\ell t_\ell(E) = 0$ implies $t_\ell(E) \cong \mathbb{Z}/\ell\mathbb{Z} \times \mathbb{Z}/\ell\mathbb{Z}$ and from there we deduce $t_{\ell^r}(E) \cong (\mathbb{Z}/\ell^r\mathbb{Z})$ for $r > 1$ and the result of the corollary in general. To prove that this kernel has m distinct points, it is sufficient to show that one fiber $n_E^{-1}(a,b)$ consists of m distinct points of E. But take for a precisely one value $a \in \mathbb{P}^1$

such that the fiber $x_n^{-1}(a)$ has $n^2 = m$ distinct solutions $a_1,\dots,a_m$. For each point a_k , there are two points on E above it, say (a_k,b_k), $(a_k,-b_k)$ (not necessarily distinct). If for one k, $b_k = -b_k$, then $b_k = 0$ for this k (we assume that $p \neq 2$) and the compatibility of n_E with the automorphism $(x,y) \longmapsto (x,-y)$ shows similarly that in $n_E(a_k,b_k) = (a,b)$ we must also have $b = 0$. This implies that all $b_k = 0$ (there must at most be m couples $(a_k,\pm b_k)$ above $(a,0)$) and that the fiber above (a,b) consists of m distinct points in all cases.

(4.19) Corollary 3. The group of points $E = E_k$ is a divisible group (when k is algebraically closed).

Proof. Because the rational mapping $x_n : \mathbf{P}^1 \longrightarrow \mathbf{P}^1$ is surjective, the diagram (4.18) shows that $n_E : E \longrightarrow E$ must be surjective (even if n is divisible by p)

Now we go to the other extreme and examine the case $n = p$. For that purpose we have to give a few prerequisites on the Lie algebra of E. Quite generally, if $\varphi : V \longrightarrow W$ is a regular morphism of algebraic varieties (defined over the algebraically closed field k), both of them having no singular point, then φ is compatible with the local rings attached to these varieties, by definition. More precisely if $P \in V$ and $Q = \varphi(P) \in W$, and if $R_P \subset k(V)$, $R_Q \subset k(W)$ denote respectively the rings of regular functions at P and Q, then $R_Q \circ \varphi \subset R_P$. We denote also by M_P the maximal ideal in R_P consisting of functions vanishing at P, so that M_P is the kernel of the evaluation mapping $f \longmapsto f(P)$ on R_P , giving an isomorphism $R_P/M_P = k$. We define the algebraic tangent space of V at P by

(4.20) $$T(V)_P = \mathrm{Der}_k(R_P,k)$$

as k-vector space. Then the regular mapping φ gives by composition a k-linear map $T(\varphi)_P = \varphi_P : T(V)_P \longrightarrow T(W)_Q$ defined by $D \longmapsto D\cdot\varphi^*$:

$$\begin{array}{ccc} f\cdot\varphi & \longleftarrow & f \\ R_P & \xleftarrow{\ \varphi^*\ } & R_Q \\ D\big\downarrow & \swarrow_{\varphi_P(D)} & \\ k & & \end{array}$$

(4.21) $\qquad (f \in R_Q \ , \ D \in \mathrm{Der}_k(R_P,k))$.

When $V = G$ is an algebraic group, the tangent space at the origin is (by definition) the Lie algebra of the group $T(G)_e = \mathrm{Lie}(G)$. The canonical injections $j_1, j_2 : G \longrightarrow G \times G$ give rise to two linear maps $\mathrm{Lie}(G) \longrightarrow \mathrm{Lie}(G \times G)$, and if we use them, it is easy to see that we can make the identification $\mathrm{Lie}(G \times G) \cong \mathrm{Lie}(G) \oplus \mathrm{Lie}(G)$. The multiplication morphism $m : G \times G \longrightarrow G$ gives rise to a k-linear map $\mu : \mathrm{Lie}(G) \oplus \mathrm{Lie}(G) \longrightarrow \mathrm{Lie}(G)$, and because $m\cdot j_1 = m\cdot j_2 = 1_G$ is the identity on G, $\mu\cdot(j_1)_e = \mu\cdot(j_2)_e = 1_{\mathrm{Lie}(G)}$ must be the identity on Lie(G). This forces μ to be the sum map of the k-vector space Lie(G). The diagonal morphism $\Delta : G \longrightarrow G \times G$, $g \longmapsto (g,g)$ has $(\Delta)_e : \mathrm{Lie}(G) \longrightarrow \mathrm{Lie}(G) \oplus \mathrm{Lie}(G)$, $D \longmapsto (D,D)$ as tangent map at the origin. Putting these facts together, we see that the homomorphism of taking the square $G \longrightarrow G$, $g \longmapsto g^2$ which is the composite $g \overset{\Delta}{\longmapsto} (g,g) \overset{m}{\longmapsto} g^2$ has $D \longmapsto (D,D) \longmapsto 2D$ as tangent map at the origin. By induction one concludes that the tangent map at the origin of $g \longmapsto g^n$ $(n \geqslant 0)$ is the multiplication by n, $n\cdot : \mathrm{Lie}(G) \longrightarrow \mathrm{Lie}(G)$ in the Lie algebra of G. If n is a multiple of the characteristic of k, the tangent map at the origin of $g \longmapsto g^n$ is thus the zero map in Lie(G). A translation argument shows that the tangent map of $g \longmapsto g^n$ at any point $x \in G$ is the zero map of $T(G)_x$ if n is divisible by the characteristic p of k.

We use these results when $G = E$ is an elliptic curve over k (hence E is a commutative algebraic group), and we still suppose that the groundfield k is algebraically closed (hence perfect).

(4.22) Proposition. If the positive integer n is divisible by the characteristic p of k, $n_E : E \longrightarrow E$ is inseparable. In particular, $\deg_s(p_E) = 1$ or p .

Proof. Write the rational function field k(E) = k(x,y) as quadratic extension of a purely transcendental extension k(x) of k, so that k(x,y) is separable over k(x) (we still assume $p \neq 2$). Then if $P \in E$, $x_P = x - x(P)$ is defined and a uniformizing parameter at P for all but finitely many P's, and $D = D_x = D_{x_P}$ for these P's. Now if $\varphi : E \longrightarrow E$, $P \longmapsto Q$ has tangent map identically 0 everywhere (which is the case for $\varphi = n_E$ when p divides n), and if we denote by $\bar{D} \in T(E)_P = \mathrm{Der}_k(R_P,k)$ the derivation $R_P \xrightarrow{D} R_P \longrightarrow k$, we shall have $(\varphi)_e(\bar{D}) = 0$. Thus, for every $f \in R_Q$, $f \circ \varphi \in R_P$ and

$$\bar{D}(f\circ\varphi) = 0 \text{ implies } D(f\circ\varphi) \in M_P \quad .$$

Let us fix $f \in k(E)$. Then $D_x(f\circ\varphi) \in \bigcap M_P$ where the intersection is taken over all $P \in E$ such that x_P is a uniformizing variable and f is regular at $Q = \varphi(P)$. As nearly all $P \in E$ satisfy these conditions, we see that the intersection in question $\bigcap M_P = \{0\}$ is reduced to 0 and $D_x(f\circ\varphi) = 0$. As $\mathrm{Der}_k(k(E))$ is a vector space over k(E) of dimension one, we conclude that $D(f\circ\varphi) = 0$ for every derivation D of k(E) (trivial on k), and that $f\circ\varphi \in k(E)^p$. Hence

$$\underbrace{k(E)\circ\varphi \subset \underbrace{k(E)^p \subset k(E)}_{\text{purely inseparable}}}_{\text{inseparable}} \quad .$$

(A more general proof could be given using results from the beginning of section (IV.1).)

(4.23) Corollary 1. The group of points $t_p(E)$ of order dividing p on E is isomorphic either to $\{1\}$ or to $\mathbb{Z}/p\mathbb{Z}$.

Proof. The proof of (4.17) above shows that only the separable degree counts for the order of the group $\mathrm{Ker}(n_E)$: $\mathrm{Card}\,\mathrm{Ker}(n_E) = \deg_s(n_E)$.

(4.24) Corollary 2. *The Tate module $T_p(E)$ is a free $\mathbb{Z}_p$-module of rank 0 or 1 and the vector space $V_p(E)$ over $\mathbb{Q}_p$ is of dimension 0 or 1.*

Proof. Observe that for two morphisms $\alpha, \beta : E \longrightarrow E$, we have $\deg(\alpha\cdot\beta) = \deg(\alpha)\cdot\deg(\beta)$, $\deg_s(\alpha\cdot\beta) = \deg_s(\alpha)\cdot\deg_s(\beta)$. In particular, $\deg_s(p_E^r) = \deg_s(p_E)^r = 1$ or p^r and

$$T_p(E) = \{1\} \text{ or } \varprojlim_r \mathbb{Z}/p^r\mathbb{Z} = \{1\} \text{ or } \mathbb{Z}_p \quad . \qquad \text{q.e.d.}$$

The two cases occur indeed, but the case $t_p(E) = \{1\}$ (or equivalently $T_p(E) = \{1\}$) only occurs when p_E is purely inseparable, $k(E)\cdot p_E = k(E)^{p^2}$, and this implies (rather easily) $E \cong E^{p^2}$, $j = j^{p^2}$, $j \in \mathbb{F}_{p^2}$. Thus, this completely degenerate case only occurs for finitely many k-isomorphism classes of curves (which can be defined over the finite field $\mathbb{F}_{p^2}$).

Using classfield theory (or rather Brauer's theory) and the natural representations of End(E) (or $\mathrm{End}_{\mathbb{Q}}(E) = \mathrm{End}(E) \otimes_{\mathbb{Z}} \mathbb{Q}$) in the ℓ-adic spaces $V_\ell(E)$ (for all primes ℓ), it can be shown that $\mathrm{End}_{\mathbb{Q}}(E)$ is of the following types

a) $\mathbb{Q}$ field of rational numbers ,

b) K a quadratic imaginary extension of $\mathbb{Q}$ in which the ideal (p) splits completely ,

c) $\mathbb{H}_{p,\infty}$ a definite quaternion algebra over $\mathbb{Q}$ with discriminant p (ramified only at p and ∞).

In the case c), End(E) must be a maximal order in $\mathbb{H}_{p,\infty}$ and this case occurs precisely when there are no points of order p on E. The finite number of invariants $j \in \mathbb{F}_{p^2}$ such that an elliptic curve with invariant j has no point of order p is thus the classnumber h_p of the quaternion algebra $\mathbb{H}_{p,\infty}$. (For further indications on this topic, look at the chapter on complements, especially (IV.1.16) and (IV.1.23)).

CHAPTER FOUR

COMPLEMENTS

As its title indicates, this last chapter provides some complements which could be considered as an introduction to the arithmetic theory of elliptic curves. In particular, the first two sections deal with properties of elliptic curves in characteristic $p \neq 0$ (or defined over finite fields) which have no conterpart in characteristic 0. Then the last section gives the first elementary results on the reduction mod p theory which associates to an elliptic curve over $\mathbb{Q}$ a family of elliptic curves over finite fields. As usual, I have not been able to refrain from mentioning some more advanced results (without proof) and some standard open conjectures.

1. Hasse's invariant

Let k be a field of characteristic $p \neq 0$ and K a finitely generated extension of k of transcendence degree one, in other words, K is a function field of one variable over k. We have shown (II.2.37) that if k is perfect ($k^p = k$), then there is a transcendental element $x \in K$ over k such that K is separable over k(x). Such elements x will be called *separating elements* for the considered extension. Quite generally, a function field K/k is called *separably generated* if it admits a separating element (and this is always the case if k is perfect).

(1.1) *Proposition*. *Let* K *be a separably generated function field* (*of one variable*) *over* k. *Then* $K = K^p \cdot k(x) = (K^p k)(x)$ *for every separating element* x *of* K. *In particular*, *if* k *is perfect*, $K = K^p(x)$ *for every separating element* x *of* K.

Proof. Let us fix a separating element $x \in K$ so that K/k(x) is a finite algebraic (separable) extension. By the primitive element theorem, there is an element $y \in K$, separable over k(x), such that $K = k(x,y)$. We have the following diagram of field extensions

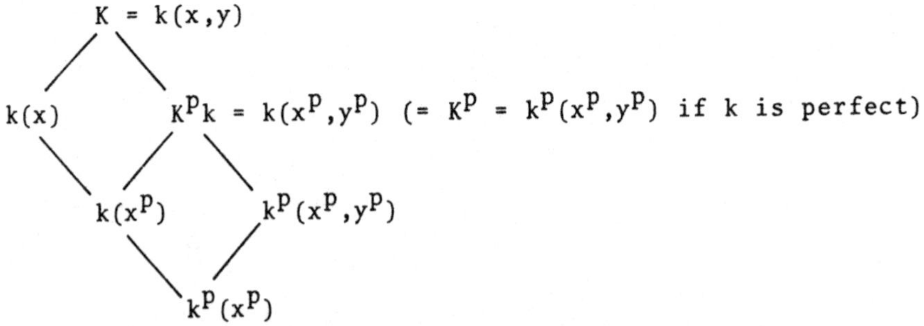

Now k(x) is (purely inseparable) of degree p over $k(x^p)$ as is immediately seen (or follows from Lüroth's theorem). If k is perfect, it follows then that the degree of K over $K^p k = K^p$ is also p, a prime

number, hence K is generated over K^p by any element not in K^p, e.g. x. The proposition is thus proved in the case k is perfect. In general, we have

$$p[K : k(x)] = [K : k(x^p)] = [K : K^p k]\,[k(x^p, y^p) : k(x^p)],$$

and we shall show that

$$[K : k(x)] = [k(x^p, y^p) : k(x^p)],$$

which will imply $[K : K^p k] = p$ and the proposition will follow as above. Let n be the degree of y over k(x) : $n = [K : k(x)]$. Then y^p is also separable of degree n over $k^p(x^p) = k(x)^p$. Because $k(x^p)$ is purely inseparable over $k^p(x^p)$, it is linearly disjoint from the separable extension $k^p(x^p, y^p)$ and this implies that the degree of y^p over $k(x^p)$ is still n which was to be proved.

(1.2) Corollary 1. Let $D \in \mathrm{Der}_k(K)$ be a derivation of K trivial over k. If $D \neq 0$, then $\mathrm{Ker}(D) = K^p k$.

Proof. If we select a separating element $x \in K$, it follows from the proposition that every element $y \in K$ can be uniquely written as

$$y = \sum_{0 \leqslant i \leqslant p-1} u_i x^i \qquad (u_i \in K^p k) .$$

Then because D must be trivial on $K^p k$,

$$D(y) = \sum_{1 \leqslant i \leqslant p-1} i u_i x^{i-1} \cdot D(x) .$$

If $D \neq 0$, also $D(x) \neq 0$ (and this is in fact equivalent to $D \neq 0$) and $D(y) = 0$ implies $u_i = 0$ for $1 \leqslant i \leqslant p-1$, hence $y = u_0 \in K^p k$. This justifies in some sense the following definition.

(1.3) Definition. The elements of $K^p k \subset K$ are called the p-constants of K whereas the other elements (i.e. elements of $K - K^p k$) are called p-variables of K.

(1.4) Corollary 2. Every p-variable of K is a separating element.

Proof. If x is any p-variable, then $K = K^p k(x)$ and the k-derivations of K are consequently fully determined by their value D(x) on x.

Because K is separably generated of transcendence degree one (and finitely generated) the space $\mathrm{Der}_k(K)$ is of dimension 1 over K, hence $D(x) \in K$ is arbitrary. This proves that every k-derivation $D \in \mathrm{Der}_k(k(x))$ has a unique extension to K, hence K is separable over $k(x)$.

(1.5) Lemma. Let K be separably generated (as before), and take two p-variables x,y of K. Then in the decomposition

$$(\tfrac{dy}{dx})/(\tfrac{y}{x}) = \sum_{0\leqslant i\leqslant p-1} u_i x^i \qquad (u_i \in K^p k),$$

the constant term u_0 is given by

$$u_0 = (\tfrac{dy}{dx})^p/(\tfrac{y}{x})^p \in K^p \subset K^p k \quad .$$

Proof. Recall that the differential quotient $\frac{dy}{dx}$ is equal to $\frac{Dy}{Dx}$ for any non-zero derivation $D \in \mathrm{Der}_k(K)$ (observe that $Dx \neq 0$ by Cor.1). Let $y = a_0 + a_1 x + \ldots + a_n x^n$ with $a_n \neq 0$, $1 \leqslant n \leqslant p-1$ be the decomposition of y relatively to x. Let K' be a decomposition field for this polynomial. As it is composite of fields of degrees dividing $n < p$, its degree must be prime to p and K' must be separable over $K^p k$. The derivation D has a unique extension as derivation D' of K'(x) over k (and D' must be 0, the only extension, on K'). Then we have

$$y = a_n \prod_{1\leqslant i\leqslant n} (x - \xi_i) \qquad (\xi_i \in K') \ .$$

Taking the logarithmic derivative, we get

$$(\tfrac{dy}{dx})/(\tfrac{y}{x}) = \sum_{1\leqslant i\leqslant n} x/(x-\xi_i) = \sum (1 - 1/(1-x\xi_i^{-1})),$$

hence

$$(\tfrac{dy}{dx})/(\tfrac{y}{x}) = \sum_{1\leqslant i\leqslant n} (1 - \frac{1}{1-x^p\xi_i^{-p}}(1+x/\xi_i+\ldots+x^{p-1}/\xi_i^{p-1})) \ .$$

The p-constant term (coefficient of x^0) is

$$u_0 = \sum_{1\leqslant i\leqslant n} (1 - 1/(1-x^p\xi_i^{-p})) = \Big\{\sum_{1\leqslant i\leqslant n} (1 - 1/(1-x\xi_i^{-1}))\Big\}^p =$$

$$= (\tfrac{dy}{dx})^p/(\tfrac{y}{x})^p \in K^p \quad .$$

(1.6) Remark. If K is separably generated over k, so is K^pk over k. Indeed, if x is a separating element for K over k, i.e. K separable over k(x), then K^p is separable over $k^p(x^p)$, and a fortiori after the purely inseparable extension $k(x^p)$ of the base field $k^p(x^p)$, we see that K^pk is separable over $k(x^p)$. This proves that x^p is a separating element of K^pk/k if x is one for K/k. If D is a k-derivation of K, we denote by D^p the k-derivation of K^pk defined by $D^p(x^p) = (D(x))^p$. Also, if $d : K \longrightarrow \mathrm{Diff}_k(K)$ denotes the canonical derivation of K, and $d^p : K^pk \longrightarrow \mathrm{Diff}_k(K^pk)$ the corresponding derivation of K^pk, then d^px^p is the linear form $D^p \longmapsto \langle D^p, d^px^p\rangle = D^p(x^p) = D(x)^p$. Because $D(x) = \langle D, dx\rangle$, we see that $\langle D^p, d^px^p\rangle = \langle D, dx\rangle^p$ which we could also write $d^px^p = (dx)^p$. In any case, if x and y are p-variables of K $(\frac{dy}{dx})^p = d^py^p/d^px^p$ is the differential quotient of the two p-variables x^p and y^p of K^pk. On the other hand, if x is a p-variable of K, and D_x denotes as usual the derivation with respect to x of K (characterized by $D_x(x) = 1$), then

$$D_x^p(x^p) = D_x(x)^p = 1^p = 1 = D_{x^p}(x^p) \text{ and } D_x^p = D_{x^p} \in \mathrm{Der}_k(K^pk) .$$

Now we turn to a more detailed study of differentials of K. By Prop. (1.1), every differential ω of K can be uniquely written

$$\omega = u_0 dx + u_1 x dx + \ldots + u_{p-1}x^{p-1}dx, \quad u_i \in K^pk$$

(once the p-variable x of K is fixed). However, it is more convenient to write it

$$\omega = f\,\frac{dx}{x} = f_0\frac{dx}{x} + f_1 dx + \ldots + f_{p-1}x^{p-2}dx \tag{1.7}$$

with p-constants $f_i \in K^pk$. Then, all terms

$$f_i x^{i-1}dx = f_i d(x^i/i) = d(f_i x^i/i)$$

are exact differentials $(1 \leqslant i \leqslant p-1)$. Consequently, the differential form ω can be written $\omega = f_0\frac{dx}{x} + \underset{\centerdot}{d}F$ with a p-constant $f_0 \in K^pk$ and an element $F \in K$ (the element f_0 being uniquely determined by x).

This shows in particular that the space $\mathrm{Diff}_k(K)/dK$ of k-differentials of K mod exact differentials of K is of dimension one over k (with basis $\frac{dx}{x}$ mod dK).

(1.8) Proposition. The operator $S_x : \mathrm{Diff}_k(K) \longrightarrow \mathrm{Diff}_k(K^p k)$ defined by $S_x(\omega) = f_0 d^p x^p / x^p$ for $\omega = f_0 dx/x + dF$ is well-defined independently of the choice of the p-variable x chosen to explicit it. This is a $K^p k$-linear operator S called Cartier operator. By definition, it vanishes on exact differentials.

Proof. Only the independence of the choice of the p-variable x deserves consideration, so let x and y be two p-variables of K. We use lemma (1.5) to compute

$$S_x(\tfrac{dy}{y}) = S_x(\tfrac{dy}{dx}/\tfrac{y}{x}.\tfrac{dx}{x}) = (\tfrac{dy}{dx})^p/(\tfrac{y}{x})^p . S_x(\tfrac{dx}{x})$$

and by the above remark (1.6) this is equal to

$$\frac{d^p y^p}{d^p x^p} / \frac{y^p}{x^p} . \frac{d^p x^p}{x^p} = \frac{d^p y^p}{y^p} = S_y(\tfrac{dy}{y}) .$$

This implies obviously $S_x = S_y$, because $\mathrm{Diff}_k(K)/dK$ is of dimension one over k. Also note that $S(\omega) = 0$ is equivalent to ω exact.

When the base field k is perfect, we can introduce the modified Cartier operator $S' : \mathrm{Diff}_k(K) \longrightarrow \mathrm{Diff}_k(K)$ by

(1.9) $$S'(f_0\frac{dx}{x} + dF) = f_0^{1/p}\frac{dx}{x} \quad (f_0 \in K^p ,\ f_0^{1/p} \in K) .$$

This operator is not linear over K^p, it is additive and satisfies the condition $S'(f^p\omega) = fS'(\omega)$ for $f^p \in K^p$ (sometimes called p^{-1}-linearity).

(1.10) Definition. A differential ω of K is called logarithmic (or logarithmically exact) if it is of the form $\frac{dx}{x}$ for some $0 \neq x \in K$.

(1.11) Proposition. Let k be a perfect field and S' the modified Cartier operator on differentials of the function field of one variable K over k. Then

a) $S'(\omega) = 0$ is equivalent to ω exact ,

b) $S'(\omega) = \omega$ <u>is equivalent to</u> ω <u>logarithmic</u> ,

c) $S'(f^p\omega + g^p\nu) = f\,S'(\omega) + g\,S'(\nu) \quad (f,g \in K,\ \omega,\nu \in \mathrm{Diff}_k(K))$.

<u>These conditions characterize uniquely the operator</u> S'.

<u>Proof</u>. Everything is obvious except perhaps b). But if $S'(\omega) = \omega$, we have $\omega = f_0 \frac{dx}{x}$ with $f_0 = f_0^{1/p}$, hence $f_0 \in \mathbb{F}_p \subset k$. Take the integer a congruent to f_0 mod p and such that $0 \leqslant a \leqslant p-1$. Then

$$\omega = f_0 \frac{dx}{x} = a \frac{dx}{x} = \frac{d(x^a)}{x^a} \quad \text{is logarithmic .}$$

It would not be difficult to see that the operator S' leaves invariant the subspace of first kind differentials of K, but we give a more precise result in our case of interest, namely the case of elliptic curves.

(1.12) <u>Theorem 1</u>. <u>Let</u> k <u>be a perfect field</u>, $1 \neq \lambda \in k^\times$, K_λ <u>the elliptic field of</u> k-<u>rational functions over the elliptic curve</u> E_λ <u>given in Legendre normal form</u> $y^2 = x(x-1)(x-\lambda)$. <u>Then the image of the first kind differential</u> $\frac{dx}{y}$ <u>over</u> E <u>under the modified Cartier operator is given by</u>

$$S'\left(\frac{dx}{y}\right) = H_p(\lambda)^{1/p} \frac{dx}{y} \quad ,$$

<u>with Deuring's polynomial</u>

$$H_p(\lambda) = (-1)^{\ell} \sum_{i=0}^{\ell} \binom{\ell}{i}^2 \lambda^i \qquad (\ell = \tfrac{1}{2}(p-1)) .$$

<u>Proof</u>. We write

$$\frac{dx}{y} = y^{-p} y^{p-1} dx = y^{-p} x^{\ell} (x-1)^{\ell} (x-\lambda)^{\ell}\, dx \ ,$$

with $\ell = \frac{1}{2}(p-1)$. Then $S'(dx/y) = (1/y) S'(x^{\ell}(x-1)^{\ell}(x-\lambda)^{\ell} dx)$.

We use the binomial expansions

$$(x-1)^{\ell} = \sum \binom{\ell}{i} x^i (-1)^{\ell - i} \quad ,$$

$$(x-\lambda)^{\ell} = \sum \binom{\ell}{j} x^j (-\lambda)^{\ell - j} \quad ,$$

and note that it is sufficient to compute the coefficient of $x^{p-1}dx$ in $x^{\ell}(x-1)^{\ell}(x-\lambda)^{\ell} dx$ because all other terms give exact differentials.

Thus we compute the coefficient of x^{ℓ} in $(x-1)^{\ell}(x-\lambda)^{\ell}$. It is

$$\sum_{i+j=\ell} \binom{\ell}{i}\binom{\ell}{j}(-1)^{i+j}\lambda^{\ell-j} = H_p(\lambda) ,$$

whence $S'(dx/y) = (1/y)S'(H_p(\lambda)x^p \frac{dx}{x}) = \frac{x}{y} H_p(\lambda)^{1/p}\frac{dx}{x}$. This gives the assertion of the theorem.

The roots of the polynomial $H_p(\lambda)$ are neither $\lambda = 0$ nor $\lambda = 1$. Indeed,

$$(-1)^{\ell} H_p(0) = 1 \neq 0 , \quad (-1)^{\ell} H_p(1) = \sum_{i=0}^{\ell} \binom{\ell}{i}^2 = \binom{2\ell}{\ell} ,$$

and $\binom{2\ell}{\ell}$ is the quotient of $(p-1)(p-2)\ldots(\frac{p+1}{2})$ by $\ell!$ having all their prime factors $< p$, hence not congruent to 0 mod p. This shows the existence of certain elliptic curves E_{λ} with $S'(dx/y) = 0$, hence with exact first kind differentials.

(1.13) Definition. An elliptic curve is called supersingular when its first kind differentials are exact.

Hence we have shown the existence of finitely many supersingular elliptic curves E_{λ} in all odd characteristics. Because the notion of first kind differential (and of exactness) only depends on the rational function field $K_{\lambda} = k(E_{\lambda})$ two values of λ leading to isomorphic curves, i.e. to the same invariant $j(\lambda)$ (as given in (I.4.3)) give simultaneously supersingular curves or not : supersingularity is a property of the invariant j of the curve. In particular, since λ and $1-\lambda$ lead to the same value for the invariant j, the two polynomials $H_p(\lambda)$ and $H_p(1-\lambda)$ (of the same degree) must have the same roots. This implies that they are proportional : $H_p(\lambda) = c_p H_p(1-\lambda)$ with a constant c_p such that $c_p^2 = 1$, hence $c_p = \pm 1 \in \mathbb{F}_p$ (note that this gives another proof of the fact that $H_p(1) \neq 0 \in \mathbb{F}_p$). Just for fun, let us compute the exact value of c_p . By definition, we have

$$c_p = H_p(1)/H_p(0) = \binom{2\ell}{\ell} = \frac{(p-1)!}{(\frac{1}{2}(p-1))!^2} ,$$

and

$$(p-1)! = 1\cdot 2\cdot 3\cdot\ldots\cdot(\tfrac{p-1}{2})(p-\tfrac{p-1}{2})(p-\tfrac{p-3}{2})\ldots(p-1)$$

is congruent mod p to

$$(\tfrac{p-1}{2})!^2(-1)^\ell \quad \text{(and also to -1 by Wilson's theorem).}$$

This proves $c_p = (-1)^\ell$ and leads to

(1.14) $$H_p(1-\lambda) = (-1)^\ell H_p(\lambda) .$$

It is obvious that

(1.15) $$H_p(1/\lambda) = (1/\lambda)^\ell H_p(\lambda) .$$

(1.16) <u>Theorem 2</u>. <u>The Hasse invariant of the family</u> E_λ <u>is given by</u>

$$H_p(\lambda) = (-1)^\ell F(\tfrac{1}{2},\tfrac{1}{2}:1:\lambda) \in \mathbb{F}_p[[\lambda]] ,$$

<u>with Gauss' hypergeometric function</u> $F = {}_2F_1$ mod p. <u>Moreover, the roots of</u> $H_p(\lambda) = 0$ <u>are all simple, and the number</u> h <u>of supersingular invariants</u> j <u>(corresponding to the roots of</u> H_p) <u>is given by the formulas</u>

$$\begin{aligned} h &= \tfrac{1}{12}(p-1) && \text{if } p \equiv 1 \bmod 12 \\ &= \tfrac{1}{12}(p-5)+1 && \quad p \equiv 5 \\ &= \tfrac{1}{12}(p-7)+1 && \quad p \equiv 7 \\ &= \tfrac{1}{12}(p-11)+2 && \quad p \equiv 11 \\ &= 1 && \text{if } p = 2 \text{ or } 3 . \end{aligned}$$

In other words, if we use the quadratic residue symbol of Legendre

$$h = \tfrac{1}{3}(1-(\tfrac{-3}{p})) + \tfrac{1}{4}(1-(\tfrac{-4}{p})) + \tfrac{1}{12}(p-1)$$

for all primes p.

<u>Proof</u> (for $p \neq 2$). Let us first check that we have the following congruence of binomial symbols

$$\binom{\frac{1}{2}(p-1)}{k} \equiv \binom{-\frac{1}{2}}{k} \quad \text{mod } p \text{ , when } p \neq 2,\ k \leqslant \tfrac{1}{2}(p-1).$$

The congruence has to be understood in $\mathbb{Z}_p$, so that mod p really means mod $p\mathbb{Z}_p$. The left-hand side is $(p-1)(p-3)\ldots(p-2k+1)/2^k k!$ while the right-hand side is $(-1)(-3)\ldots(-2k+1)/2^k k!$. Because the denominators $2^k k!$ are units in $\mathbb{Z}_p$ for $p \neq 2$, the asserted congruence amounts to the trivial congruence

$$(p-1)(p-3)\dots(p-2k+1) - (-1)(-3)\dots(-2k+1) \equiv 0 \quad (p) .$$

In fact the binomial coefficients $\binom{-\frac{1}{2}}{k}$ are rational numbers with powers of 2 only in their denominators (look at the proof of (II.5.28)) $\binom{-\frac{1}{2}}{k} \in \mathbb{Z}[\frac{1}{2}] \subset \mathbb{Z}_p$ (p prime $\neq 2$), and these numbers are in $p\mathbb{Z}_p$ when $k > \frac{1}{2}(p-1)$ so that we have

$$(-1)^{\ell} H_p(\lambda) = \sum_{k=0}^{\ell} \binom{\ell}{k}^2 \lambda^k = \sum_{k=0}^{\infty} \binom{-\frac{1}{2}}{k}^2 \lambda^k = F(\tfrac{1}{2},\tfrac{1}{2}:1:\lambda) \in \mathbb{F}_p[[\lambda]] .$$

But this hypergeometric function satisfies the differential equation (hypergeometric differential equation with $a = b = \frac{1}{2}$, $c = 1$)

$$\lambda(1-\lambda)F'' + (1-2\lambda)F' - \tfrac{1}{4}F = 0,$$

Where we consider this equation as differential equation for formal series. This proves that the roots of this function are simple ($F(\lambda) = F'(\lambda) = 0$ implies $F''(\lambda) = 0$ because we have already checked that the roots are neither 1 nor 0, and differentiating, we would get $F'''(\lambda) = 0,\dots$, hence finally $F = 0$ as formal series). It only remains to prove the formula for the number of supersingular j's. I recall that (I.4.3)

$$j = 2^8 \lambda^{-2}(1-\lambda)^{-2}(1-\lambda+\lambda^2)^3$$

(and this formula <u>defines</u> j in characteristic p = 3), and thus the affine λ-line is a covering of the affine j-line (Lüroth's theorem) of degree 6, with ramified points over $j = 0$ of ramification indices 3 corresponding to the roots λ_i $(i = 1,2)$ of $1 - \lambda + \lambda^2 = 0$, and over $j = 12^3$ with ramification indices 2 at the points $\lambda = -1, \frac{1}{2}, 2$. Here we have implicitely assumed $p \neq 3$ (and $p \neq 2$) because if $p = 3$, $12^3 \equiv 0 \bmod 3$ shows that all ramification points collapse over $j = 0$ and give an index of ramification 6 at $\lambda = -1$ over $j = 0 = 12^3 \in \mathbb{F}_3$. Before giving the general discussion, let us treat the case of characteristic $p = 3$. Then H_3 is of degree one, has only one root and there is only one supersingular invariant j. But all values of λ corresponding to this value of j must also be roots, so that the only possibility is $\lambda = -1$ of ramification index 6 above $j = 0 = 12^3$. Indeed, $H_3(\lambda) = -1 - \lambda = 0$ also gives $\lambda = -1$! Consider now the case $p > 3$ along similar lines. Suppose that there is a prime p such that $H_p = 0$ does <u>not</u> have roots corresponding to $j = 0$ or 12^3 (I do not claim that such p exist, but it will be a consequence of the discussion that they do exist!). Then, for each value of supersingular j, there will be six values of λ above j, and we shall have the formula $\ell = \frac{1}{2}(p-1) = 6h$, or $p = 12h + 1$ and this shows that p would necessarily be congruent to 1 mod 12 in this case. Suppose now that p is such that $j = 0$ is supersingular but $j = 12^3$ is not supersingular in characteristic p. Then the h-1 supersingular values of j distinct from 0 must come from six roots λ of H_p and above $j = 0$ there will be only two corresponding values for λ :

$$\ell = \tfrac{1}{2}(p-1) = 6(h-1) + 2 \quad .$$

In this case, we would thus have

$$p = 12(h-1) + 5 \equiv 5 \bmod 12.$$

Similarly, if p is such that $j = 0$ and 12^3 are supersingular in characteristic p, we get

$$\ell = \tfrac{1}{2}(p-1) = 6(h-2) + 2 + 3 \quad ,$$

hence

$$p = 12(h-2) + 11 \equiv -1 \bmod 12 \ .$$

Finally suppose that $j = 12^3$ is supersingular but $j = 0$ is not, in characteristic p, then

$$\ell = \tfrac{1}{2}(p-1) = 6(h-1) + 3$$

leads to $\qquad p = 12(h-1) + 7 \equiv -5 \bmod 12$.

Because these different conditions for p are mutually exclusive, they must characterize the different cases and all occur (if we use Dirichlet's theorem about primes in arithmetic progressions, we see that all these cases occur infinitely many times - with the same density). Moreover, we have found the following formulas (where Σ_p denotes the finite set of supersingular invariants j in characteristic p) :

$$\begin{aligned} h = \mathrm{Card}(\Sigma_p) &= \tfrac{1}{12}(p-1) \text{ if } p \equiv 1 \ (12) \text{ case } \Sigma_p \not\ni 0, 12^3 , \\ &= \tfrac{1}{12}(p+7) \text{ if } p \equiv 5 \quad \text{"} \quad 0 \in \Sigma_p , 12^3 \notin \Sigma_p , \\ &= \tfrac{1}{12}(p+13) \text{ if } p \equiv -1 \quad \text{"} \quad 0, 12^3 \in \Sigma_p \\ &= \tfrac{1}{12}(p+5) \text{ if } p \equiv -5 \quad \text{"} \quad 0 \notin \Sigma_p , 12^3 \in \Sigma_p . \end{aligned}$$

This concludes the proof of the theorem, giving more precisely the cases when 0 or 12^3 are supersingular (in characteristic 2, there is only one supersingular curve, namely $y^2 + y = x^3$, and with a suitable definition for its invariant - given by Deuring - it gives $\Sigma_2 = \{0\}$).

Although we have already used the term "Hasse's invariant" of the curve E_λ as being the value $H_p(\lambda)$ of Deuring's polynomial, let us give a formal definition now for this terminology. First, we observe that by definition, $S'(\frac{dx}{y}) = H_p(\lambda)^{1/p} \frac{dx}{y}$ where S' denotes the modified Cartier operator, hence also

(1.17) $$S(\tfrac{dx}{y}) = H_p(\lambda)(\tfrac{dx}{y})^p \quad .$$

In general, if E is an elliptic curve given with a first kind differential $\omega \neq 0$, we define the Hasse invariant $H(E,\omega)$ by

(1.18) $$S(\omega) = H(E,\omega)\,\omega^p \quad .$$

If we replace ω by any other first kind differential, say $\omega' = a\omega$ with $a \in k^\times$, then obviously $H(E,a\omega) = a^{1-p}H(E,\omega)$, so that we can say that the Hasse invariant is defined up to multiplication by elements of the form a^{1-p} independently of the choice of the basic first kind differential. When the elliptic curve is given by an equation $y^2 = f(x)$ with a cubic polynomial f having all its roots distinct, the choice of first kind differential $\omega = \frac{dx}{y}$ is sometimes implicitely made, and we have

(1.19) $H(E,\frac{dx}{y}) = H(E) = \underline{\text{coefficient of}}\ x^{p-1}\ \underline{\text{in}}\ f(x)^{\frac{1}{2}(p-1)}$

as in the proof of Th.1 (1.12).

Let us take now E in Weierstrass form $y^2 = 4x^3 - g_2x - g_3$ and let us compute its Hasse invariant (with respect to $\frac{dx}{y}$) in function of g_2 and g_3. We have to find the coefficient of $x^{2\ell}$ ($\ell = \frac{1}{2}(p-1)$ as before) in $(4x^3 - g_2x - g_3)^{\ell}$. Using twice the binomial expansion formula in a suitable way we get

$$(4x^3 - g_2x - g_3)^{\ell} = \sum_{m=0}^{\ell} \sum_{n=0}^{\ell-m} \frac{\ell!}{m!n!(\ell-m-n)!} 4^{\ell-m-n}(-g_2)^m(-g_3)^n x^{3\ell-2m-3n}$$

and because $3n = \ell - 2m$ implies $n \leqslant \ell - m$, the required coefficient is

(1.20) $$H(E) = \sum_{2m+3n=\ell} (-1)^{m+n} 4^{\ell-m-n} \frac{\ell!}{m!\,n!\,(\ell-m-n)!} g_2^m g_3^n \quad .$$

This is an isobaric polynomial of degree ℓ in g_2 and g_3 if these elements are given the respective weights 2 and 3. On the other hand, the Eisenstein series E_{ℓ} can also be expressed as isobaric polynomial of the same form in g_2 and g_3, but coefficients in characteristic 0 (I.3.10). The result is that

(1.21) $H(g_2,g_3) = (E_{\ell}(g_2,g_3) \bmod p)$ (Deligne) .

Let us just sketch a proof of this formula. We compare the q-expansions of both sides. By (I.4.1)

$$E_k(z) = 1 + \gamma_k \sum_{n \geqslant 1} \sigma_{2k-1}(n)q^n \quad (q = e^{2\pi iz} = \underline{e}(z)) ,$$

with

$$\gamma_k = (-1)^k 4k/B_k \quad (B_1 = 1/6,\ B_2 = 1/30,\ B_3 = 1/42,\dots).$$

But the denominator of B_k is the product of the primes m such that m-1 divides 2k (von Staudt), so that γ_{ℓ} is always divisible by p : The q-expansion of $(E_{\ell} \bmod p)$ is identically 1. We show that the same is true for the q-expansion of $H(g_2,g_3)$. To find this q-expansion, we evaluate the value of the Hasse invariant of Tate's curve $K^{\times}/q^{\mathbb{Z}}$ with the local field (of characteristic p) $\mathbb{F}_p((q)) = K$ (the additive valuation of this field is defined by $\operatorname{ord}(q) = 1$, hence $|q| = \frac{1}{p} < 1$). The differential $\frac{dx}{y}$ takes the form

$$\frac{dx}{y} = \frac{dP}{DP} = \frac{dP}{(1/X)(d/dX)(P)} = \frac{dX}{X} \quad .$$

Formally (this has to be justified), this is a logarithmically exact differential, hence $S(\frac{dx}{y}) = (\frac{dx}{y})^p$ and $H = 1$ also. This shows that the two q-expansions in question coïncide and (1.21) follows from this.

We mention a few related results without proof. First let E be an elliptic curve defined over the finite field $\mathbb{F}_q$ of characteristic p (say $q = p^f$). Then the number N of rational points of E (including the point at infinity) with coordinates in $\mathbb{F}_q$ satisfies the following congruence

(1.22) $$N \equiv 1 - H^{(q-1)/(p-1)} \mod p \qquad {}^{*)} \quad .$$

Here again, H denotes the Hasse invariant of E, with respect to any $\mathbb{F}_q$-rational first kind differential: if $a \in \mathbb{F}_q^{\times}$

$$(a^{1-p}H)^{(q-1)/(p-1)} = H^{(q-1)/(p-1)} \in \mathbb{F}_p \quad .$$

Suppose now that E is supersingular, i.e. $H = 0$. Then the above congruence, applied to all finite extensions $\mathbb{F}_{q'}$ of $\mathbb{F}_q$ in a fixed algebraic closure $\bar{\mathbb{F}}_p$ shows that $E(\bar{\mathbb{F}}_p)$ has no p-torsion. A consequence of (1.22) is thus that the supersingular elliptic curves have no point of division of order p. Conversely, if $H \neq 0$, the norm of H in an extension $\mathbb{F}_{q^m}$ of $\mathbb{F}_p$ being the mth power of the norm of H relatively to the extension $\mathbb{F}_q/\mathbb{F}_p$ will be 1 for large m, and this will show that E has division points of order p (rational over

---

*) Observe that $H^{(q-1)/(p-1)} = H^{p^{f-1}+\dots+1} = \operatorname{Norm}_{\mathbb{F}_q/\mathbb{F}_p}(H) \in \mathbb{F}_p$.

the algebraic closure of $\mathbb{F}_p$). In fact one can prove the following equivalences :

(1.23)
i) The Hasse invariant $H(E)$ is zero ,

ii) $E(\bar{\mathbb{F}}_p)$ has no point of order p ,

iii) The ring of endomorphisms $\mathrm{End}(E)$ is not commutative,

iv) A power π^n of the Frobenius $(x \longmapsto x^q)$ of E is equal to an integer (necessarily $p^{n/2}\cdot 1_E$) .

Thus any one of these properties characterizes supersingular elliptic curves. Moreover, if E is supersingular, multiplication by p gives an isomorphism $p\cdot 1_E$ purely inseparable of degree p^2, hence E is isomorphic to its p^2 power $E \approx E^{p^2}$. Consequently $j = j(E) = j^{p^2}$ and this shows that all supersingular invariants are in $\mathbb{F}_{p^2}$.

(1.24) Remark. As in characteristic 0, we say that an elliptic curve in characteristic p is singular, or that its invariant is singular (to avoid confusion with singular points) when its ring of endomorphisms is different from $\mathbb{Z}$: $\mathrm{End}(E) \supsetneq \mathbb{Z}$. The singular invariants are exactly the elements $j \in \overline{\mathbb{F}}_p$. When moreover $\mathrm{End}(E)$ is not commutative, the corresponding singular invariant is supersingular. This explains the terminology.

(1.25) Curiosity. We may observe the following formulas using the classical polynomials of Legendre P_ℓ . First we have

$$P_{2i}(0) = \binom{-\frac{1}{2}}{i} \quad .$$

Also

$$(1-\lambda)^\ell P_\ell\left(\frac{1+\lambda}{1-\lambda}\right) = \sum_{i=0}^{\ell} \binom{\ell}{i}^2 \lambda^i \quad ,$$

so that the Hasse invariant of Legendre's family is given by

$$H_p(\lambda) = (\lambda-1)^\ell P_\ell\left(\frac{1+\lambda}{1-\lambda}\right) \bmod p \qquad (\ell = \tfrac{1}{2}(p-1)\ ) \quad .$$

2. Zeta function of an elliptic curve over a finite field

Let V be a non-singular, irreducible, projective plane curve defined over the finite field $k = \mathbb{F}_q$ of characteristic $p \neq 0$ ($q = p^f$) and let K be the function field k(V) of k-rational functions on V. (Actually, we could start with any function field K of one variable over $\mathbb{F}_q$, but it may be more suggestive to speak of points of the curve V than of places of the field K.) As usual V denotes the set of points in an algebraic closure $\bar{k}$ of k : $V = V(\bar{k}) \subset \mathbb{P}^2(\bar{k})$ and Div(V) is the free abelian group generated by the points of V (finite formal combinations of points of V with integral coefficients). The subgroup of rational divisors $\mathrm{Div}_k(V) \subset \mathrm{Div}(V)$ consists of divisors $\underline{d} = \sum d_P(P)$ having equal multiplicities on conjugate points $d_P = d_{P^\sigma}$ (for every automorphism $\sigma \in \mathrm{Gal}(\bar{k}/k)$). We define <u>the norm of a divisor</u> $\underline{d}$ by the formula

$$N(\underline{d}) = q^{\deg(\underline{d})} \quad ,$$

and the <u>zeta function of</u> V by the formal Dirichlet series

(2.1)
$$\zeta_V(s) = \sum_{\underline{d} \geqslant 0} N(\underline{d})^{-s} \quad .$$

We shall see that this series converges absolutely for $\mathrm{Re}(s) > 1$, but we treat it formally now.

A positive divisor $\underline{d} \in \mathrm{Div}_k(V)$ is called <u>prime</u> (rational over k) if it cannot be expressed as the sum of two positive non-zero divisors. Obviously prime rational divisors are of the form

$\mathfrak{p} = \sum (P^\sigma)$ sum over a class of conjugate elements of V.

The degree of such a prime rational divisor is the number of conjugates of any point occurring in it, and if k(P) is the field generated by the coordinates of P (one of them being normalized to 1) over k, this degree must be the number of automorphisms of k(P)/k, hence

$\deg(\mathfrak{p}) = [k(P) : k]$. Because every positive divisor can be expressed uniquely as sum of prime divisors (rational over k), we see that (2.1) is equivalent to

(2.2) $$\zeta_V(s) = \prod_{\mathfrak{p} \text{ prime}} (1 - N(\mathfrak{p})^{-s})^{-1}$$

product extended to the (positive) prime divisors over V, rational over k. If we take $u = q^{-s}$ as variable, this formula reads

(2.3) $$Z_V(u) = \zeta_V(s) = \prod_{\mathfrak{p} \text{ prime}} (1 - u^{\deg(\mathfrak{p})})^{-1} .$$

Let us take the logarithms of both sides (I recall that we are making formal computations so far)

$$\begin{aligned} \log Z_V(u) &= \sum_{\mathfrak{p}} \log(1 - u^{\deg(\mathfrak{p})})^{-1} = \\ &= \sum_{\mathfrak{p}} \sum_{m\geqslant 1} u^{m \deg(\mathfrak{p})}/m = \sum_{\mathfrak{p}} \deg(\mathfrak{p}) \sum_{m\geqslant 1} \frac{u^{m \deg(\mathfrak{p})}}{m \deg(\mathfrak{p})} \\ &= \sum_{n\geqslant 1} \Big\{ \sum_{\deg(\mathfrak{p})\mid n} \deg(\mathfrak{p}) \Big\} u^n/n . \end{aligned}$$

But if $\mathfrak{p} = \sum (P^{\sigma})$, the condition $\deg(\mathfrak{p}) \mid n$ means that $[\mathbb{F}_q(P) : \mathbb{F}_q] \mid n$, or that $k(P) \subset \mathbb{F}_{q^n}$, and is equivalent to $P \in V(\mathbb{F}_{q^n})$. If we recall then that $\deg(\mathfrak{p})$ is precisely the number of distinct conjugates of P, we obtain the expression

(2.4) $$\log Z_V(u) = \sum_{n\geqslant 1} N_n u^n/n ,$$

with $$N_n = \operatorname{Card} V(\mathbb{F}_{q^n}) = \sum_{\deg(\mathfrak{p})\mid n} \deg(\mathfrak{p}) .$$

Two equivalent formulas are

(2.5) $$Z'_V(u)/Z_V(u) = \sum_{n\geqslant 1} N_n u^{n-1} \quad (\log Z_V(0) = 0) ,$$

(2.6) $$Z_V(u) = \exp \sum_{n\geqslant 1} N_n u^n/n .$$

Because

$$N_n = \operatorname{Card} V(\mathbb{F}_{q^n}) \leqslant \operatorname{Card} \mathbb{P}^2(\mathbb{F}_{q^n}) = q^{2n} + q^n + 1 ,$$

we see that the series (2.4-6) converge absolutely for $|u| < q^{-2}$ and thus for $\operatorname{Re}(s) > 2$, thereby proving the convergence of (2.1) for large $\operatorname{Re}(s)$ and the legitimacy of the formal computations in that domain. Also, if V is a straight line, we see from (2.6) that

$$Z_{line}(u) = \frac{1}{(1-u)(1-qu)} \quad ,$$

because

$$\mathrm{Card}\ \mathbb{P}^1(\mathbb{F}_{q^n}) = q^n + 1 \quad .$$

Thus the zeta function of a line has a meromorphic extension to the complex s-plane as rational function in q^{-s}. This fact is general, and one can prove that the zeta function of a curve of genus g has the form

(2.7) $$Z_V(u) = \frac{P_{2g}(u)}{(1-u)(1-qu)}$$

with a polynomial $P_{2g}(u) \in \mathbb{Z}[u]$ of degree 2g in u, satisfying

$$P_{2g}(0) = 1\ ,\ P_{2g}(1) = N_1\ ,\ P_{2g}(u) = q^g u^{2g} P_{2g}(1/qu) \quad .$$

This implies the functional equation

$$\zeta_V(1-s) = q^{(1-2s)(1-g)}\ \zeta_V(s) \quad .$$

A. Weil has also proved that the zeros of these zeta functions all lie on the critical line $\mathrm{Re}(s) = \frac{1}{2}$ (for g = 1, this was due to H. Hasse). The main interest of the explicit knowledge of the zeta function of a curve as rational function in u resides in the fact that it is equivalent to the knowledge of all number of points N_n of V (over the extension of degree n of $k = \mathbb{F}_q$). Actually, using the functional equation, we see that the determination of the g coefficients of $u, \ldots, u^g$ in the polynomial P_{2g} suffices to determine completely the zeta function (and hence <u>all</u> N_n for $n \geq 1$). For an elliptic curve, the rationality of Z_V is very easy to show.

(2.8) <u>Theorem</u>. <u>Let</u> E <u>be an elliptic curve defined over</u> $k = \mathbb{F}_q$ <u>(i.e. an absolutely non-singular projective plane cubic over</u> $\mathbb{F}_q$ <u>with one rational point chosen as neutral element</u>). <u>The zeta function</u> $\zeta_E(s)$ <u>extends as meromorphic function of s in the whole complex plane and satisfies</u> $\zeta_E(1-s) = \zeta_E(s)$. <u>This function is rational in</u> $u = q^{-s}$

$$\zeta_E(s) = Z_E(u) = \frac{1-(1+q-N_1)u + qu^2}{(1-u)(1-qu)} \quad ,$$

<u>with</u> $N_1 = \mathrm{Card}\ E(\mathbb{F}_q) > 0$.

Proof. The function field k(E) of k-rational functions over E is of genus one (II.2.26) (when $p \neq 2$, which we will suppose for this proof; because we assume that there is a rational point $P \in E(\mathbb{F}_q)$, the proof of the reference given is still valid although k is not algebraically closed). Let $\mathrm{Div}_k^n(E)$ denote the set of k-rational divisors on E of degree $n \geqslant 0$, and $\mathrm{Div}_k^n(E)/P(E)$ the set of classes of divisors of degree n mod principal divisors (this has a meaning because the principal divisors have degree zero). From Abel-Jacobi's condition, we see that this set is parametrized by the rational points of E (the sum of the affixes of the points in a rational divisor over k is a point of E which must be equal to all its conjugates, hence be in $E_k = E(\mathbb{F}_q)$), and in particular has a finite number N of elements equal to the number of rational points N_1 on E (and in particular independent of the degree n in Div^n). On the other hand, it is easy to compute the number of positive divisors in a given class (of degree $n > 0$). For $\underline{d} \in \mathrm{Div}_k^n(E)$ and $f \in k(E)$,

$$\underline{d} + \mathrm{div}(f) \succcurlyeq 0 \iff \underline{d} \succcurlyeq -\mathrm{div}(f) \iff f \in L(\underline{d})$$

and thus the number of positive divisors in the class of $\underline{d}$ is the number of principal divisors div(f) with $f \in L(\underline{d})$, $f \neq 0$. Since a function is determined up to a multiplicative constant by its divisor, this number is

$$\frac{q^{\dim L(\underline{d})} - 1}{q - 1} .$$

Because the field k(E) is of genus one, $\dim L(\underline{d}) = \deg(\underline{d})$ if $\deg(\underline{d}) > 0$. This shows that

$$\sum_{\underline{d} \succ 0} N(\underline{d})^{-s} = \sum_{d=\deg \underline{d} \geqslant 1} q^{-ds} N \frac{q^d - 1}{q - 1} =$$

$$= \frac{N}{q-1} \sum_{d \geqslant 1} (q^{(1-s)d} - q^{-sd}) =$$

$$= \frac{N}{q-1}(q^{1-s}/(1-q^{1-s}) - q^{-s}/(1-q^{-s})) .$$

Hence putting $u = q^{-s}$ we get (adding the term 1 corresponding to $\underline{d} = 0$)

$$Z_E(u) = 1 + \frac{N}{q-1}(qu/(1-qu) - u/(1-u)) =$$

$$= \frac{1 - (q+1-N)u + qu^2}{(1-u)(1-qu)} ,$$

which is the desired expression. One checks immediately the functional equation on this "explicit" formula. On this explicit formula we see that the zeta function Z_E converges for $|u| < 1/q$ or for

$$|q^{-s}| < 1/q , \quad q^{\mathrm{Re}(s)} = |q^s| > q , \quad \mathrm{Re}(s) > 1$$

(we had only checked the convergence for $|u| < q^{-2}$, $\mathrm{Re}(s) > 2$ after (2.6)) .

There is a formula for the number of rational points for an elliptic curve over $\mathbb{F}_p$ given in Legendre form

$$y^2 = x(x-1)(x-\lambda) \qquad (\lambda \in \mathbb{F}_p) .$$

For each finite $x \in \mathbb{F}_p$, there will be no point, one point, two points on $E(\mathbb{F}_p)$ with first coordinate x if $x(x-1)(x-\lambda)$ is resp. not a square, zero, a square of $\mathbb{F}_p$. Using the quadratic residue symbol, this number is $(\frac{x(x-1)(x-\lambda)}{p}) + 1$, and if we add the point at infinity, we find (in this case)

(2.9) $$N = 1 + \sum_{x \in \mathbb{F}_p} \left\{ (\frac{x(x-1)(x-\lambda)}{p}) + 1 \right\} .$$

(We use the convention $(\frac{a}{p}) = 0$ when $a = 0$ for the quadratic residue symbol.)

(2.10) Remark. If we write the numerator of the zeta function of the elliptic curve E in the form

(2.1) $$1 - cu + qu^2 = (1 - \omega_1 u)(1 - \omega_2 u) ,$$

Hasse has shown first that the inverse roots ω_i have absolute value $q^{\frac{1}{2}}$. This proves that the zeros of this zeta function are all on the critical line Re(s) = $\frac{1}{2}$. The corresponding general property for the zeta functions of curves (any genus) over finite fields has been established by A. Weil (see the references concerning this chapter).

Finally there is result connecting the zeta functions of two isogenous curves : if E and E' are two elliptic curves defined over $\mathbf{F}_q$, and if there exists a non-zero homomorphism E $\longrightarrow$ E' (defined over some extension of $\mathbf{F}_q$), then the zeta functions of E and of E' over $\mathbf{F}_q$ are the same.

3. Reduction mod p of rational elliptic curves

Let k be a number field (finite algebraic extension of $\mathbb{Q}$) and $\mathfrak{p}$ a (non-zero) prime ideal of the ring of integers of k. By elliptic curve over k, we always mean a non-singular (over $\bar{k}$) projective plane cubic defined by a polynomial with coefficients in k, with one rational point over k chosen as neutral element for the group law (usually, this neutral rational element will be on the line at infinity with a suitable coodinate system in the projective plane).

(3.1) Definition. The elliptic curve E over k is said to have good reduction mod $\mathfrak{p}$ if there exists a suitable coordinate system in the projective plane $\mathbb{P}^2$ in which E is given by an equation with coefficients in the ring of $\mathfrak{p}$-integers $\mathfrak{o}_{\mathfrak{p}}$ of k, this equation mod $\mathfrak{p}$ still defining an elliptic curve (non-singularity condition) over the residual finite field $\mathbb{F}_q = \mathfrak{o}_k/\mathfrak{p} = \mathfrak{o}_{\mathfrak{p}}/\mathfrak{p}\mathfrak{o}_{\mathfrak{p}}$ $(q = N_{k/\mathbb{Q}}(\mathfrak{p}))$.

We shall always assume that the prime ideal $\mathfrak{p}$ is not a divisor of 6 (hence q prime to 2 and 3), so as to be able to work with Weierstrass equations. If E has good reduction mod $\mathfrak{p}$, the reduced curve has thus a Weierstrass equation, and the original curve can be chosen in Weierstrass form with a discriminant $\Delta \in \mathfrak{o}_{\mathfrak{p}}$ with $\mathrm{ord}_{\mathfrak{p}}\Delta = 0$ (because the non-singularity condition for the reduced Weierstrass equation is precisely $(\Delta \bmod \mathfrak{p}) = \tilde{\Delta} \neq 0$). If we take any Weierstrass equation for E over k, its discriminant will be in k with $\mathrm{ord}_{\mathfrak{p}}\Delta = 0$ for nearly all (i.e. all but a finite number including the divisors of 2 and 3) prime ideals of $\mathfrak{o}_k$, hence :

(3.2) Proposition 1. Every elliptic curve over k has good reduction mod $\mathfrak{p}$ for nearly all prime ideals (not dividing 2 and 3) of the ring of integers of k.

(3.3) Proposition 2. Let $\mathfrak{p}$ be a prime ideal (not dividing 2 or 3) of the number field k, and E an elliptic curve over k. If E has good reduction mod $\mathfrak{p}$, its absolute invariant $j = j(E)$ satisfies the following conditions :

a) $\mathrm{ord}_{\mathfrak{p}} j \geqslant 0$, $\mathrm{ord}_{\mathfrak{p}} j \equiv 0 \bmod 3$,

b) $\mathrm{ord}_{\mathfrak{p}}(j - 12^3) \equiv 0 \bmod 2$.

Conversely, if $j \in k$ satisfies these conditions, there exists an elliptic curve over k with invariant j having good reduction mod $\mathfrak{p}$.

Proof. If E has good reduction mod $\mathfrak{p}$, it has a Weierstrass equation $y^2 = 4x^3 - g_2 x - g_3$ with $\mathrm{ord}_{\mathfrak{p}} g_i \geqslant 0$ and $\mathrm{ord}_{\mathfrak{p}} \Delta \neq 0$. This implies that its invariant $j = 12^3 g_2^3/\Delta$ satisfies $\mathrm{ord}_{\mathfrak{p}} j = 3\mathrm{ord}_{\mathfrak{p}}(12 g_2) \geqslant 0$ (divisible by 3), and similarly for $j - 12^3 = 2^6 3^6 g_3^2/\Delta$, $\mathrm{ord}_{\mathfrak{p}}(j - 12^3) =$ $= 2\mathrm{ord}_{\mathfrak{p}}(2^3 3^3 g_3)$ must be even (and positive). Conversely, we assume that j satisfies the enounced conditions. The cases $j = 0$ and $j = 12^3$ are easily treated separately, so we assume that j does not take these two special values. Define then $g = 27j/(j-12^3)$ ($\neq 27$, $\neq 0$ because $j \neq 0$), so that the elliptic curve $y^2 = 4x^3 - gx - g$ has invariant j. We have $g - 27 = 27 \cdot 12^3/(j - 12^3)$, and the discriminant of this curve is $g^2(g - 27)$. For simplicity write $v = \mathrm{ord}_{\mathfrak{p}}$. Then $v(g^2(g - 27)) =$ $= 2v(g) + v(g - 27) = 2v(j) - 3v(j - 12^3)$ must be a multiple of 6 by hypothesis. Choose $a \in k$ such that $v(a^6 g^2(g - 27)) = 0$ and put $g_2 = a^2 g$, $g_3 = a^3 g$. Then the elliptic curve $y^2 = 4x^3 - g_2 x - g_3$ has still the invariant j, has a discriminant Δ satisfying $v(\Delta) = 0$ and $0 \leqslant v(j) = 3v(12 g_2)$ implies $v(g_2) \geqslant 0$. Finally, since $v(\Delta) = 0$ and using $v(g_2) \geqslant 0$, we get $v(g_3) \geqslant 0$.

Observe that this proposition does not mean that any elliptic curve over k with invariant j satisfying the conditions has good reduction mod $\mathfrak{p}$, but it is isomorphic over $\bar{k}$ to an elliptic curve over k with good reduction (another k-form of the given elliptic curve).

The preceding proposition shows that the condition $\mathrm{ord}_{\mathfrak{p}} j \geqslant 0$ is not sufficient to imply the good reduction mod $\mathfrak{p}$ of an elliptic curve over k of invariant j. However :

(3.4) Corollary. Let E be an elliptic curve over k with invariant j, $\mathrm{ord}_{\mathfrak{p}} j \geqslant 0$. Then there exists a finite algebraic extension k'/k such that when considered as elliptic curve over k', E has good reduction mod $\mathfrak{P}$ for all prime ideals (of $\mathfrak{o}_{k'}$) dividing $\mathfrak{p}$. (This is the potential good reduction of E at $\mathfrak{p}$ as defined by J.-P. Serre and J. Tate.)

Proof. It is sufficient to take for k' a normal extension ramified at $\mathfrak{p}$ with indices $e(\mathfrak{P}|\mathfrak{p})$ (independent of $\mathfrak{P}$ dividing $\mathfrak{p}$) multiple of 6. Then all conditions of the proposition will be satisfied.

Let S be a finite set of prime ideals of the ring of integers of k. We denote by $\mathfrak{o}_{(S)} = \mathfrak{o}[S]^{-1}$ the subring of k consisting of elements such that $\mathrm{ord}_{\mathfrak{p}} x \geqslant 0$ for all prime ideals $\mathfrak{p}$ not belonging to S (fractions having denominators in S). If S is given, we can always enlarge it in a finite set $S' \supset S$ such that $\mathfrak{o}_{(S')}$ is a principal ideal ring. This follows from the fact that the ideal class group of k is finite combined with the characterization of ideals in a ring of fractions (if $\mathfrak{a}_1, \ldots, \mathfrak{a}_h$ are representing ideals for the class group of k, it is sufficient to take for S' any finite set of ideals containing S and the prime divisors of the $\mathfrak{a}_i$'s).

(3.5) Proposition 3. Let S be a finite set of prime ideals of $\mathfrak{o}_k$ such that $\mathfrak{o}_{(S)}$ is principal, and let E be an elliptic curve over k. E has good reduction mod $\mathfrak{p}$ for all prime $\mathfrak{p} \notin S$ if and only if E has a Weierstrass equation with coefficients $g_i \in \mathfrak{o}_{(S)}$ and $\Delta = g_2^3 - 27 g_3^2 \in \mathfrak{o}_{(S)}^{\times}$.

Proof. That the conditions are sufficient is obvious. Conversely, we have to establish the existence of one Weierstrass equation with simultaneous good reduction outside S. Let $y^2 = 4x^3 - g_2' x - g_3'$ be any Weierstrass equation for E over k. If $\mathfrak{p}$ is a prime not in S, there exists by assumption, a Weierstrass equation for E over k, of the form $y^2 = 4x^3 - g_2(\mathfrak{p})x - g_3(\mathfrak{p})$ with $g_i(\mathfrak{p}) \in \mathfrak{O}_{\mathfrak{p}}$ and $\Delta(\mathfrak{p}) \in \mathfrak{O}_{\mathfrak{p}}^{\times}$. Necessarily $g_2(\mathfrak{p}) = g_2' \cdot a^4(\mathfrak{p})$, $g_3(\mathfrak{p}) = g_3' \cdot a^6(\mathfrak{p})$ and $\Delta(\mathfrak{p}) = \Delta' \cdot a^{12}(\mathfrak{p})$ with some $a(\mathfrak{p}) \in k$. Because the first model (with the primes) has good reduction at nearly all primes (not in S) of k (Prop.1), we may choose $g_i(\mathfrak{p}) = g_i'$ $(i = 2,3)$ and $a(\mathfrak{p}) = 1$ for nearly all $\mathfrak{p} \notin S$. Now the ideal generated by $\prod_{\mathfrak{p} \notin S} \mathfrak{p}^{\mathrm{ord}_{\mathfrak{p}} a(\mathfrak{p})}$ in $\mathfrak{O}_{(S)}$ must be principal, and this proves the existence of an element $a \in k$ with

$$\mathrm{ord}_{\mathfrak{p}} a = \mathrm{ord}_{\mathfrak{p}} a(\mathfrak{p}) \text{ for all } \mathfrak{p} \notin S .$$

The transformation of coordinates $x \mapsto a^{-2}x$, $y \mapsto a^{-3}y$ will then give the equation $y^2 = 4x^3 - g_2x - g_3$ for E over k, with $g_2 = a^4 g_2' \in \mathfrak{O}_{(S)}$, $g_3 = a^6 g_3' \in \mathfrak{O}_{(S)}$ and $\Delta = a^{12}\Delta' \in \mathfrak{O}_{(S)}^{\times}$ as desired.

(3.6) <u>Theorem</u> (<u>Šafarevič</u>). <u>Let</u> S <u>be any finite set of prime ideals of</u> $\mathfrak{O}_k$. <u>Then the set of</u> k-<u>isomorphism classes of elliptic curves with good reduction outside</u> S <u>is finite</u>.

<u>Proof</u> (modulo a theorem of Siegel). Let us enlarge S until it contains all divisors of 2 and 3, and $\mathfrak{O}_{(S)}$ is principal. By proposition 3, any elliptic curve over k with good reduction outside S has a Weierstrass equation $y^2 = 4x^3 - g_2x - g_3$ with $g_i \in \mathfrak{O}_{(S)}$ and $\Delta \in \mathfrak{O}_{(S)}^{\times}$. Now Δ is well-defined mod $(\mathfrak{O}_{(S)}^{\times})^{12}$, and by Dirichlet's unit theorem, $\mathfrak{O}_{(S)}^{\times}$ is finitely generated - more precisely isomorphic to the product of a finite abelian group by a free abelian group of rank $\mathrm{Card}(\bar{S}) - 1$, where $\bar{S}$ denotes the union of S with all the archimedian places of k - so that $\mathfrak{O}_{(S)}^{\times}/(\mathfrak{O}_{(S)}^{\times})^{12}$ is <u>finite</u>. It is thus sufficient to prove the theorem when the discriminant Δ is <u>fixed</u>. But $\xi^3 - 27\eta^2 = \Delta$ is the equation of an elliptic curve

$$\eta^2 = (\xi/3)^3 - \Delta/27 \quad (\text{invariant } j = 0).$$

We are looking for the points on this elliptic curve, having coordinates $\xi = g_2$ and $\eta = g_3$ in the ring $\mathcal{O}_{(S)}$. Siegel's theorem asserts that there are only finitely many such points.

(3.7) Remark. Siegel originally proved that on any affine curve of genus $g \geqslant 1$, there are only finitely many points with integral coordinates. Mahler and Lang proved an improved version of this theorem replacing the ring of integers $\mathcal{O}_k$ by rings of the form $\mathcal{O}_{(S)}$ (S as above). The full proof of Siegel's theorem uses the Mordell-Weil theorem through Jacobians, hence is quite difficult. However, (3.6) only uses the result of Siegel in the case of an elliptic curve ($g = 1$) of absolute invariant $j = 0$ (essentially, only Mordell's part, for this curve, is needed).

Let E be an elliptic curve over k, S a finite set of prime ideals of $\mathcal{O}_k$ (containing the divisors of 2 and 3) such that E has good reduction outside S. If $\mathfrak{p} \notin S$, the reduced curve $E_{\mathfrak{p}}$ is an elliptic curve over the finite field $\mathcal{O}_k/\mathfrak{p}$ and the numerator of the zeta function of $E_{\mathfrak{p}}$ is

$$1 - a_{\mathfrak{p}} N(\mathfrak{p})^{-s} + N(\mathfrak{p})^{1-2s} \quad \text{with } a_{\mathfrak{p}} = 1 + N(\mathfrak{p}) - \mathrm{Card}\, E_{\mathfrak{p}}(\mathcal{O}/\mathfrak{p}).$$

The Hasse-Weil zeta function of E is defined (up to a finite number of factors) by the infinite product

$$\prod_{\mathfrak{p} \notin S} (1 - a_{\mathfrak{p}} N(\mathfrak{p})^{-s} + N(\mathfrak{p})^{1-2s})^{-1} \quad . \tag{3.8}$$

This infinite product is known to converge absolutely for large Re(s) and defines a holomorphic function in a right half-plane. There are several standard conjectures (verified for special curves and for elliptic curves admitting complex multiplications) concerning the meromorphic extension of this zeta function to the whole plane, and a functional equation under the transformation $s \longmapsto 2 - s$. It is also conjectured (Birch and Swinnerton-Dyer) that the order of the zero at $s = 1$ of this function is precisely the rank of the group

E(k) of rational points over k. Remember that by the Mordell-Weil theorem, this abelian group is finitely generated, hence product of a finite group (torsion subgroup) by a free abelian group $\mathbb{Z}^r$ with a rank r defined by $r = \dim_{\mathbb{Q}} E(k) \otimes \mathbb{Q}$.

(3.9) Remark. To obtain the simplest functional equation possible, it is necessary to define all factors of the zeta function of E (not only those for $\mathfrak{p} \notin S$ where E has good reduction). For $\mathfrak{p}$ (not dividing 6), we take a Weierstrass equation of E over $k_{\mathfrak{p}}$ with $\mathfrak{p}$-integral coefficients $g_i(\mathfrak{p})$ and discriminant $\Delta(\mathfrak{p})$ of minimal possible order. Thus $0 \leqslant \mathrm{ord}_{\mathfrak{p}} \Delta(\mathfrak{p}) < 12$. Bad reduction occurs precisely when $\mathrm{ord}_{\mathfrak{p}} \Delta(\mathfrak{p}) > 0$ and then the reduced curve mod $\mathfrak{p}$ is an irreducible cubic (in Weierstrass form) with either a node (two roots only of the cubic polynomial mod $\mathfrak{p}$ coïncide, leading to a double point) or a cusp (the three roots of the cubic polynomial mod $\mathfrak{p}$ coïncide). The addition law of E reduces in the first case in a multiplicative group on the set of regular points, and in the second case, it reduces to an additive group on the set of regular points (cf. (II.2.5) and (II.2.6)). Thus the nodal case is said to correspond to multiplicative reduction, and the cuspidal case to additive reduction. The Euler factor attached to such a place with bad reduction has then the form

(3.10) $$(1 - \delta_{\mathfrak{p}} N(\mathfrak{p})^{-s})^{-1}$$

with $\delta_{\mathfrak{p}} = \pm 1$ in the nodal case (the +1 being chosen when the tangents at the node are rational over $\mathcal{O}_k/\mathfrak{p}$, i.e. when the two points in the normalized curve sitting above the node are rational over that field), and $\delta_{\mathfrak{p}} = 0$ in the cuspidal case.

A prime $\mathfrak{p}$ giving cuspidal reduction is more degenerate than a point with nodal reduction, and in fact the cuspidal reduction can be avoided by enlarging the groudfield k (which cannot be done with the nodal or multiplicative reduction : this is why this reduction is also called semi-stable).

(3.11) Proposition. Let E be an elliptic curve defined over the number field k, and let $\mathfrak{p}$ be a non-zero prime ideal (not dividing 6) of the ring of integers $\mathcal{O}_k$ of k. Then there exists a finite extension k' of k such that additive reduction never occurs for ideals $\mathfrak{p}'$ in k' dividing $\mathfrak{p}$.

Proof. Let $y^2 = 4x^3 - g_2x - g_3$ be a Weierstrass equation of E over k. Let also e_i denote the three distinct roots of the right-hand side (in an algebraic closure $\bar{k}$ of k) and define $k' = k(\sqrt{e_i})_{i=1,2,3}$. Over k', we can write the equation of E in the form

$$y^2 = 4(x - e_1)(x - e_2)(x - e_3) ,$$

with $e_1 + e_2 + e_3 = 0$. Choose (and fix) a prime divisor $\mathfrak{p}'$ of $\mathfrak{p}$ in k', and use $v = \mathrm{ord}_{\mathfrak{p}'}$ for the corresponding valuation. Using a transformation $x \longmapsto \lambda^2 x$, $y \longmapsto \lambda^3 y$ with $\lambda = 1/\sqrt{e_i}$ (for a suitable i), we are led to the case where

$$e_i = 1 \ , \ v(e_j) \geqslant 0 \ (j=1,2,3) .$$

If the e_j were all congruent mod $\mathfrak{p}'$ to $a \in \mathcal{O}_{k'}/\mathfrak{p}'$, we would have

$$0 = \sum e_j \equiv 3a \bmod \mathfrak{p}' \quad \text{hence} \quad a = 0 \in \mathcal{O}_{k'}/\mathfrak{p}' ,$$

which leads to a contradiction to the fact that $1 = e_i \equiv 1 \not\equiv 0 \bmod \mathfrak{p}'$. Thus at least two of the reduced roots mod $\mathfrak{p}'$ are distinct and the cuspidal case cannot occur.

The definition (and this proposition) shows that the zeta function of an elliptic curve E depends on the field k over which E is considered to be defined.

(3.12) <u>Further results</u>. Two elliptic curves defined over some algebraic number fields are called <u>isogenous</u> when their corresponding lattices in $\mathbb{C}$ are isogenous (III.3.8) or equivalently, when there exists a non-zero homomorphism from one to the other (this homomorphism will then necessarily be surjective with finite kernel, and be defined over an algebraic closure of the number fields in question). Then it can be proved that two isogenous elliptic curves have the same (Hasse-Weil) zeta functions*. Conversely, it is <u>conjectured</u> that two elliptic curves (defined over number fields) with the same zeta function are isogenous. This converse has been proved for curves with <u>non-integral</u> invariants j (look at Th. (III.3.10)) by J.-P. Serre .

---

*) Over a common number field of definition.

REFERENCES

FOR CHAPTER I

Let us start with some representative classics:

BRIOT C., BOUQUET C. : Théorie des fonctions elliptiques, 2e éd. Paris, Gauthiers-Villars 1875, 2 vol.

FRICKE R. : Die elliptischen Funktionen und ihre Anwendungen, Leipzig-Berlin, Teubner 1922.

APPEL P., LACOUR E. : Principes de la théorie des fonctions elliptiques et applications, 2e éd. Paris, Gauthier -Villars 1922.

This last book should especially be consulted for applications such as spherical pendulum, plane elastic curve, heat theory,...

Among more recent books on function theory containing a chapter on elliptic functions, we quote only

WHITTAKER E.T., WATSON G.N. : (A course of) Modern analysis, Cambridge at the University Press, 4th ed. reprinted 1965.

But a reader interested in the most simple properties of elliptic functions only should start with

AHLFORS L.V. : Complex analysis, 2d ed. New-York, McGraw-Hill, 1966.

For the theory of theta functions, we have adopted notations close to those of Weil in

WEIL A. : Introduction à l'étude des variétés kählériennes, Act. Sc. et Industrielles 1267, Paris, Hermann 1958.

Let us also mention

WEIL A. : Théorèmes fondamentaux de la théorie des fonctions thêta, Séminaire Bourbaki Mai 1949.

SIEGEL C.L. : *Vorlesungen über gewählte Kapitel der Funktionentheorie* (Notes by Gottschling E., Klingen H.), Mathematische Institut der Universität, Göttingen, 1964.

For automorphic functions and forms,

FORD L.R. : *Automorphic functions*, 2^d ed. New-York, Chelsea Publ. 1951.

LEHNER J. : *Discontinuous groups and automorphic functions*, (Math. Survey 8) Amer. Math. Soc. Providence, 1964.

Shorter introductions are provided by

GUNNING R.C. : *Lectures on modular forms* (Notes by Brumer A.), Princeton University Press, Princeton N.J., 1962.

SERRE J.-P. : *Cours d'arithmétique* (Collection "Le Mathématicien"), Presses Universitaires de France, Paris 1970.

The formula $SL_2(\mathbb{Z})/(\pm 1) = \mathbb{Z}/(2) * \mathbb{Z}/(3)$ and interesting generalizations can be found in

SERRE J.-P. : *Arbres, amalgames et* SL_2 , (notes rédigées avec la collaboration de Bass H.), to appear in the Springer-Verlag lecture notes in mathematics series, Berlin.

I have selected Siegel's method for the derivation of the infinite product of Δ (Jacobi's theorem) :

SIEGEL C.L. : *A simple proof of* $\eta(-1/\tau) = \eta(\tau)\sqrt{\tau/i}$ Mathematika 1, 1954, p.4 (Complete works, vol 2, p.188).

SIEGEL C.L. : *Analytische Zahlentheorie* (Notes by Kürten K.F., Köhler G.), Mathematische Institut der Universität, Göttingen, 1964.

For more formulas connecting elliptic functions to theta functions

ERDELYI-MAGNUS-OBERHETTINGER-TRICOMI : *Higher Transcendental Functions*, (Bateman Manuscript Project), New-York, McGraw-Hill, 1953,vol.2.

For Chapter II

Prerequisites on noetherian rings, integers, extension of valuations (and the "Nullstellensatz") are all contained in

LANG S. : Algebra, Addison-Wesley Publ.Co., Reading Mass. Third printing 1971 (World Student Series ed.).

For a good introduction to the theory of algebraic curves (Bezout and Riemann-Roch's theorem) on a relatively elementary level, a student could start with

FULTON W. : Algebraic curves, Benjamin Inc., New-York, 1969.

One can also look at

CHEVALLEY C. : Introduction to the theory of algebraic functions of one variable (Survey 6) Amer. Math. Soc. Providence, 1951.

EICHLER M. : Einführung in die Theorie der algebraischen Zahlen und Funktionen, Basel u. Stuttgart, Birkhäuser, 1963.

These books contain also an introduction to the algebraic theory of elliptic functions. For the classification of elliptic differential forms and integrals, one can consult

COURANT R. HUREWITZ A. : Funktionentheorie, (Die Grundlehren...Bd.3) Berlin, Springer-Verlag, 1922.

For analytic p-adic function theory, the basic material is contained in

DWORK B. : On the zeta function of a hypersurface, I.H.E.S. Publ. Math. No 12 , Bures-sur-Yvette, 1962.

GUNTZER A. : Zur Funktionentheorie einer veränderlichen über einem vollständigen nichtarchimedischen Grundkörper, Archiv d. Math., 17, 1966, pp.415-431.

Note however that in this last article, the groundfield is always supposed to be algebraically closed and complete, so that a little bit more work has to be done to get Schnirelmann's theorem in its strongest form (II.4.16), or its corollary (II.4.18). Schnirelmann's original reference is

SCHNIRELMANN L. : Sur les fonctions dans les corps normés et algébriquement fermés, Bull. Acad. Sci. USSR. Math, vol. 5, pp.487-497.

Tate's p-adic elliptic curves are treated in

ROQUETTE P. : Analytic theory of elliptic functions over local fields, Göttingen, Vandenhoeck and Ruprecht, 1970.

A sketch of this theory is also given in

SERRE J.-P. : Abelian ℓ-adic representations and elliptic curves, New-York, Benjamin Inc., 1968.

For Chapter III

Division points on abelian varieties are treated in

LANG S. : Abelian varieties, Interscience Publ. Inc., New-York, 1959.

MUMFORD D. : Abelian varieties, Oxford University Press (for the Tata Institute of fundamental Research, Bombay), Oxford, 1970.

A more elementary point of view is adopted by

SHIMURA G. : Arithmetic theory of automorphic functions (Introduction to the...), Iwanami Shoten, Publ., and Princeton University Press, 1971,

who gives the classical proof for the integrality of singular invariants. We have also followed this book for the presentation of the second section. More on ℓ-adic representations will be found in

SERRE J.-P. : Abelian ℓ-adic representations and elliptic curves, New-York, Benjamin Inc., 1968,

already referred to for chapter II.

An introduction to the basic ideas of etale cohomology (for pedestrians...) is given in the talk

MUMFORD D. : Arithmetical algebraic geometry, Proceedings of a Conference held at Purdue University, ed. by O.F.G. Schilling, Harper & Row, Publ., New-York, 1965, pp. 33-81.

For the finiteness theorem for the rational torsion, we have given Weil's proof as indicated by J.W.S. Cassels in the excellent survey article

CASSELS J.W.S. : Diophantine equations with special reference to elliptic curves, Journal London Math. Soc.,41, 1966, pp.193-291.

For the original proof, see

LUTZ E. : Les solutions de l'équation $y^2 = x^3 - Ax - B$ dans les corps p-adiques, C.R.Acad.Sc. Paris, 203,1936,pp.20-22.

WEIL A. : Sur les fonctions elliptiques p-adiques, loc.cit. pp.22-24.

Let us also quote

SERRE J.-P. : p-torsion des courbes elliptiques, Sém. Bourbaki, vol. 1969-70, exposé 380, pp.281-294.

Plenty of references on elliptic curves are given in the survey article by Cassels, and in the book on ℓ-adic representations by Serre. For more recent references, one can look at the bibliography of the book by Shimura and that of

SERRE J.-P. : Propriétés galoisiennes des points d'ordre fini des courbes elliptiques, Inv. Math., 15, 1972, pp. 259-331.

For Chapter IV

The Cartier operator is defined and studied in Eichler's book (reference given for Chapter II). A standard reference for the Hasse invariant of elliptic curves is

DEURING M. : Die Typen der Multiplikatorenringe elliptischer Funktionenkörper, Abh. Math. Sem. Univ. Hamburg, 14, 1941, pp.197-272.

We have chosen Igusa's method to prove that the roots of $H_p(\lambda) = 0$ are simple and to determine the number of supersingular invariants.

IGUSA J.I. : Class number of a definite quaternion with prime discriminant, Proc. Nat. Acad. Sc. USA, 44, 1958, pp.312-314 .

Another way of defining the zeta function of a curve over a finite field, using an adelic integral, is given in

WEIL A. : Basic Number Theory, (Die Grundlehren ... Bd.144) , Berlin, Springer-Verlag, 1967 .

Observe that he gives the functional equation with the explicit rational form of this function (Th.4, p.130), but omits to prove that the coefficients of the polynomial in the numerator are integers. For a proof that there exists only finitely many integral solutions on the curve $y^2 = x^3 + k$, look at

MORDELL L.J. : Diophantine equations, London - New-York, Academic Press, 1969.

Since I have omitted to define the absolute invariant j of an elliptic curve in characteristic 2, the interested reader will have to go back to the appendix of Roquette's book on p-adic elliptic functions (reference given for Chapter II), where he will find reduced forms for all characteristics. The original article is

DEURING M. : Invarianten und Normalformen elliptischer Funktionenkörper, Math. Zeitschrift, 47, 1941, pp. 47-56.

For finer reduction properties of elliptic curves, we refer to the third chapter of

NERON A. : Modèles minimaux des variétés abéliennes sur les corps locaux et globaux, I.H.E.S. Publ. Math. 21, Bures-sur-Yvette, 1964, .

and for applications of minimal models to

SERRE J.-P., TATE J. : Good reduction of abelian varieties, Ann. of Math., 88, 1968, pp. 492-517.

The Riemann hypothesis for the zeta function of elliptic curves over finite fields is proved in the book by S. Lang on abelian varieties. One can also come back to

WEIL A. : Sur les courbes algébriques et les variétés qui s'en déduisent, Act. Sc. et Industrielles 1041 , Paris, Hermann 1948.

INDEX *)

---

*) References are to pages : II.27 refers to page 27 of chapter II .